OKANAGAN UNIV/COLLEGE LIBRARY
02649663

D0787319

DATE DUE
BRODART
Cat. No. 23-221

Global Warming and the Built Environment

Professional Development Foundation

The Professional Development Foundation is a non-profit research trust limited by guarantee and registered in England. Together with the European Professional Development Foundation, its sister organization based in Italy, the PDF specializes in the promotion of practitioner generated research.

Through a network of advisers in higher education, and public and private sector organizations, the PDF provides consultancy, research support and professional development programmes. Both independent projects and cooperative schemes with other professional bodies and academic institutions are undertaken. Publications and research projects have covered Environmental Health, Disaster Management and Prevention, Health and Safety, Transportation Management, European Management Development, Law and Psychology, Personnel Management, Solid Waste Management, European Community Issues, Child and Adolescent Therapy, Community Development, Bullying in Schools and Persistent Deliquency. In the early 1970s the Foundation pioneered competence-based approaches to professional development, linking the workplace, professional bodies and universities.

The Professional Development Foundation targets neglected areas of research in order to raise professional awareness. Interest in environmental concerns started in this way with research into eduction in environmental health, submitted as a development proposal to the Royal Institute of Public Health and Hygiene (1969), and in our development of the health care counselling schemes with the Royal Society of Health (1978).

Solarch

Solarch is a multi-disciplinary research, teaching and consultancy unit based at the University of New South Wales. Its expertise includes passive solar design, energy efficiency, standards development and testing for buildings and materials, the ecological impact of the built environment and design responsibility, urban design, building user patterns and preferences, pre-design research and design management and post-occupancy evaluation.

Solarch is active in information dissemination, public education programmes and professional development courses. Its publications include energy efficiency and architectural design issues, guidelines on passive heating and natural cooling, solar access and site planning, public housing and consumer attitudes and experiences.

The Foundation prepared a series of recommendations in 1972 (the United Nations Year of the Environment) focusing on environmental health and the need to develop a coherent approach to education at the school, community and professional level. It was decided to review progress in key areas in the succeeding twenty years. The Foundation's advisers, Robert Samuels and David A. Lane, produced a proposal on environment health, and Robert Samuels also proposed a review of the issues of global warming and the built environment. The editors of this book, Robert Samuels and Deo K. Prasad, produced the final proposal. As members of the National Solar Architecture Research Unit (Solarch) they were ideally placed to implement the proposal. Solarch and the Professional Development Foundation adopt an holistic approach to environmental issues and this is the prevailing ideology developed in this book.

The Professional Development Foundation and Solarch are pleased to present *Global Warming and the Built Environment* as a valuable addition to the literature.

Professional Development Foundation
21 Limehouse Cut
46 Morris Road
London
E14 6NQ
England

Solarch
The University of New South Wales
School of Architecture
Box 1, Kensington
Sydney, NSW 2052
Australia

Global Warming and the Built Environment

Edited by

Robert Samuels

Senior Research Fellow, Solarch; Senior Lecturer, School of Architecture, University of New South Wales, Sydney, Australia

and

Deo K. Prasad

Director, Solarch; Senior Lecturer, School of Architecture, University of New South Wales, Sydney, Australia

E & FN SPON
An Imprint of Chapman & Hall
London · Glasgow · Weinheim · New York · Tokyo · Melbourne · Madras

Published by E & FN Spon, an imprint of Chapman & Hall, 2–6 Boundary Row, London SE1 8HN, UK

Chapman & Hall, 2–6 Boundary Row, London SE1 8HN, UK

Blackie Academic & Professional, Wester Cleddens Road, Bishopbriggs, Glasgow G64 2NZ, UK

Chapman & Hall GmbH, Pappelallee 3, 69469, Weinheim, Germany

Chapman & Hall Inc., One Penn Plaza, 41st Floor, New York, NY 10119, USA

Chapman & Hall Japan, Thomson Publishing Japan, Hirakawacho Nemoto Building, 6F, 1–7–11 Hirakawa-cho, Chiyoda-ku, Tokyo 102, Japan

Chapman & Hall Australia, Thomas Nelson Australia, 102 Dodds Street, South Melbourne, Victoria 3205, Australia

Chapman & Hall India, R. Seshadri, 32 Second Main Road, CIT East, Madras 600 035, India

First edition 1994

Phototypeset in 10/12pt Palatino by Intype, London
Printed in Great Britain by St Edmundsbury Press, Bury St Edmunds

ISBN 0 419 19210 7

A catalogue record for this book is available from the British Library

Library of Congress Catalog Card Number: 93–74896

∞ Printed on permanent acid-free text paper, manufactured in accordance with ANSI/NISO Z39.48–1992 and ANSI/NISO Z39.48–1984 (Permanence of Paper).

This book is dedicated
to the children
of the 21st century

Contents

Contributors

George Baird
Senior Lecturer,
School of Architecture,
Victoria University of Wellington,
New Zealand

John A. Ballinger
Professor and Head,
School of Architecture,
University of New South Wales,
Sydney, Australia

Deborah Cassell
Architect and Research Associate,
Solarch, School of Architecture,
University of New South Wales,
Sydney, Australia

Bente L. Christensen
Krusemyntegade 24,
DK 1318,
Copenhagen K,
Denmark

Jeffrey Cook
Regents Professor,
College of Architecture and
Environmental Design,
Arizona State University, USA

Paul R. Ehrlich
Bing Professor of Population
Studies,
Department of Biological Sciences,
Stanford University, USA

David A. Lane
Director,
Professional Development
Foundation,
21 Limehouse Cut, 46 Morris
Road, London, UK;
Visiting Professor,
Syracuse University, USA;
Honorary Senior Fellow,
City University and University
College, London, UK

Ian Lowe
Associate Professor and Director,
Science Policy Research Centre,
Department of Science and Technology,
Griffith University, Queensland, Australia

Julius Malkin
Director,
Professional Development Foundation,
21 Limehouse Cut, 46 Morris Road, London, UK;
Fellow,
Regents College,
City University, London, UK

Peter Newman
Associate Professor and Director,
Institute for Science and Technology Policy,
Murdoch University,
Perth, Australia

Jorgen S. Norgard
Energy Group,
Physics Department,
Building 309,
Technical University of Denmark,
DK-2800 Lyngby, Denmark

Patrick O'Sullivan OBE
Professor and Head,
Bartlett School of Architecture and Planning, University College,
London, UK

John Page
Emeritus Professor,
ex-Cambridge Interdisciplinary Environmental Centre,
Department of Geography,
University of Cambridge, UK

Deo K. Prasad
Director,
Solarch,
and Senior Lecturer,
School of Architecture, University of New South Wales, Sydney,
Australia

Allan Rodger
Professor,
Department of Architecture and Building,
University of Melbourne,
Australia

Robert Samuels
Senior Research Fellow,
Solarch,
and Senior Lecturer,
School of Architecture,
University of New South Wales,
Sydney, Australia

Gerald Vinten
Whitbread Professor of Business Policy,
Luton University College of Higher Education,
Luton, UK

Foreword
Uncertainty and insurance

Paul R. Ehrlich

There is no question that, by altering the surface of the Earth and adding heat trapping gases to the atmosphere, the expanding human enterprise is changing the climate. But the degree and speed of alteration, the direction and nature of regional and local changes, and the consequences of change, are extremely difficult to predict. The situation is especially difficult because many of the things that humanity must do to support its burgeoning numbers will further exacerbate climate change. These include burning fossil fuels (which add carbon dioxide to the atmosphere) to supply additional people with power, expanding rice agriculture to feed growing numbers (which releases another greenhouse gas, methane), using more fertilizer to increase crop yields (releasing heat-trapping nitrous oxide), and building more houses, factories and other facilities, and clearing more land (changing the reflectivity of the planetary surface, adding more greenhouse gases).

The risks that humanity face should it be unlucky in the 'lottery' of climate change are gigantic. Rapid climate change could seriously damage an already threatened agricultural enterprise (Daily and Ehrlich, 1990; Ehrlich *et al.*, 1993; Parry, 1990), bringing starvation to hundreds of millions and the spectre of plague and social breakdown. The frequency of catastrophic hurricanes could greatly increase, and if sea levels rise as much as some predict entire island nations could disappear.

Part of the uncertainty comes from ignorance of natural systems. For example, the climate is not well enough understood, computer models not sufficiently detailed and available computer time is inadequate to permit regional climatic changes to be predicted with any certainty. Another substantial part of the uncertainty comes from ignorance of human behaviour. We simply do not know what people are going to do. To what degree will solar-hydrogen or other systems that do not release carbon dioxide be substituted for fossil fuels? How rapidly will natural gas substitute for coal in various uses? Will the Montreal ozone protocol be strengthened and well enforced? How much will flooded rice production be expanded? What will be the pace of desertification, deforestation? Will the rate of population growth continue to slow or will growth resurge? Will the built environment continue to expand, destroying more agricultural ecosystems and natural ecosystems?

In the face of such uncertainty, what strategy should humanity adopt? I believe the answer is straightforward. Since the potential threat is so large, it would clearly behove us to take out substantial insurance. The chances of climatic catastrophe are in the vicinity of 50% if most climatologists are correct. We are quite accustomed to insuring against risks in the vicinity of 1% or less, as when a young person takes out term life insurance or anyone insures their home against fire or earthquake.

To put it another way, we should adopt a 'fail-safe' approach to the design of spaceship Earth. When a new passenger aircraft is designed, several (usually four) separate hydraulic systems are included, each independently able to control the aircraft in case of multiple failures. Few of us would voluntarily ride on an aircraft that had two of its four systems inoperable. Yet the political approach to global environmental problems has generally been to keep pressing the ecological systems that support civilization as if to see how much damage it will take to bring on a general collapse. We need to treat the only place in the universe known to be capable of supporting life more conservatively. We must hedge against eco-catastrophe.

Taking out insurance against climatic disaster means taking all reasonable steps to slow the flux of greenhouse gases into the atmosphere and to limit land-use changes that influence the climate (Ehrlich and Ehrlich, 1991). Energy conservation should be a prime goal in all nations. Coal as an energy source should be phased out as rapidly as possible, and more climate-benign technologies phased in – such as employing solar cells to produce electricity and using some of the electricity to produce hydrogen for use as a portable fuel for automobiles and aircraft. Agricultural practices should be changed as rapidly as possible to minimize the flow of methane and nitrous oxides into the atmosphere from farm fields. Chlorofluorocarbon manufacture should be phased out as rapidly as possible. Deforestation should be halted and replaced by programmes of reforestation. And, of course, the growth of the human population should be stopped as soon as humanely possible, and a slow decline begin (Ehrlich and Ehrlich, 1990, 1991).

While efforts are made to slow the pace of change, society should also strive to increase its ability to cope with the change that cannot be averted. Flexibility must be the rule in agriculture. Contingency plans should be made for substituting new strains of crops (or new crops) as farm fields are subjected to novel environmental conditions, and crop genetic research targeted more strongly on drought-resistance (since the drying of breadbasket areas would be the most likely adverse climatic change). Plans should be made for strengthening agricultural extension services, and food storage and famine relief capabilities. Coastal zone development (which should be limited in any case) should include planning for both sea-level rise and increased frequency and intensity of storms. Public health should begin planning for a migration of tropical diseases.

The degree to which these and other steps should be taken is, of course, a matter of social decision. The basic question is 'How much is one willing to insure against catastrophe?'. Fortunately, most of the steps that should be taken in the built environment fall into the general category of 'no regrets' measures. Increasing the energy and durability of buildings will reduce emissions of carbon dioxide from heating/cooling and from mining and manufacturing. Making vehicles more efficient will do the same. So will gradually rebuilding cities so that they are no longer designed primarily around automobiles rather than people. People should at worst be able to commute on efficient mass transport systems, and at best be able to walk or bicycle to work.

Such steps would be desirable even if there were no prospect of a greenhouse catastrophe. Costs of heating, cooling and commuting would drop. Damage to human health and infrastructure from air pollution would be reduced. If walking to work became the rule the health of the population would be improved because of the benefits of increased exercise.

The basic problem on the planet today is that the scale of the human enterprise has become so large that it can only be maintained by eroding the health of society's life-support systems (Ehrlich and Ehrlich, 1990, 1991). For example, that the entire planet is overpopulated can be seen from the simple standard that humanity cannot live on its 'income' but only by continuing to consume a one-time inheritance of natural capital – especially deep, rich, agricultural soil, fossil groundwater and biodiversity.

Biodiversity, the plants, animals and microorganisms with which we share the earth, is the resource that is most irreplaceable and simultaneously most threatened by the rapid climatic change likely to result from greenhouse warming (Wyman, 1991). Those living organisms are working parts of the ecosystems that supply the human economy with essential and irreplaceable services (Ehrlich and Ehrlich, 1981). These services include maintenance of a benign mix of gases in the atmosphere, for instance, not too much or too little of those that create the greenhouse effect; running the hydrolic cycle that supplies fresh water to society; generating and maintaining the soils essential to agriculture and forestry and keeping those soils fertile; controlling the vast majority of potential pests that would attack our crops and vectors of human disease, pollinating many crops, and maintaining a vast genetic library from which the very basis of civilization has already been withdrawn in the form of crops, domestic animals, medicines and industrial products.

In short, an insurance strategy for dealing with the potential effects of global warming involves reducing the three factors that multiply together to generate the impact of *Homo sapiens* on earth's ecosystems: the size of the population, the average level of affluence (or consumption) and the environmental damage (including contributing to the warming) done by the technologies employed to supply each unit of consumption. Architects, engineers, geographers, city planners and all others concerned with the built environment can contribute a great deal towards developing a fail-safe approach to our 'spaceship'. One way they must do it, ironically, is by limiting sharply the scale of the built environment.

References

Daily, G. C. and Ehrlich, P. R. (1990) An exploratory model of the impact of rapid climate change on the world food situation. *Proc. R. Soc. Lond. B.*, **241**, 232–44.

Ehrlich, P. R. and Ehrlich, A. H. (1981) *Extinction: The Causes and Consequences of the Disappearance of Species*, Random House, New York.

Ehrlich, P. R. and Ehrlich, A. H. (1990) *The Population Explosion*, Simon and Schuster, New York.

Ehrlich, P. R. and Ehrlich, A. H. (1991) *Healing the Planet*, Addison Wesley, New York.

Ehrlich, P. R., Ehrlich, A. H. and Daily, G. C. (1993) Food security, population, and environment. *Population and Development Review*, **19**, 1–32.

Parry, M. (1990) *Climate Change and World Agriculture*, Earthscan, London.

Wyman, R. L. (ed.) (1991) *Global Climate Change and Life on Earth*, Routledge, Chapman & Hall, New York.

Introduction
Global warming and the built environment – the challenge

David A. Lane and Julius Malkin

We are responsible

The idea that we are responsible for our environment is currently gaining favour in national and international circles. As citizens we are faced with a confusing array of conflicting opinion. Is global warming real? Does it matter anyway?

We would argue that it does matter and there are a number of important steps that can be taken. As citizens our impact is real; on policy through pressure groups and the ballot box; and through our own behaviour as purchasers of professional services and consumers of scarce resources. We make a difference.

As members of professions that contribute to the built environment we have a particularly important part to play. Unfortunately, we are faced with conflicting demands and a literature which is scattered and difficult to interpret. This book is an attempt to make that literature more accessible and thereby inform responsible professional debates and practice. The perspective adopted by the editors (Robert Samuels and Deo Prasad) is a holistic one, focused on an attempt to understand the interaction between the different elements that make up the built environment. If we are to achieve the aim of a sustainable approach to development which is ecologically responsible, then we must understand the complex interactions between the built and natural environment. It is our belief that the built environment contributes significantly to global warming. Understanding that relationship, in particular, the core role of energy use and its impact, does generate agendas for action which are both achievable and sustainable.

The work of leading contributors is presented to give an international view of the built environment. In this introduction we review the concerns they raise and present some of their arguments. They provide evaluations of the complex interactions that we must grasp and tackle in order to foster a significant change in direction in our attitudes to the environment that we create and destroy. The book starts with the challenge that faces us and considers the extent to which we are responsible. We must be prepared to undertake careful audits of our contributions and establish options for change to the extent that we are accountable.

In looking at the picture presented, the global perspective is considered through evaluation of the vital contribution of the human use of renewable energy. This is the repetitive theme of the book. The relationship between global warming and urban development is examined and the question whether we have the foresight to face the enormity of the task confronting us asked. This question leads us to the realization that urban design has not featured strongly in the greenhouse debate and yet how we build, and the interaction of city planning and transportation systems, are major factors in the production of greenhouse gases. These broad questions provide the basis for the case in favour of sustainable development policies.

The major strategies are examined broadly and then in detail in terms of the specific options open to us. These options provide a discussion of an approach to architectural form in terms of climate modification and the development of a benign architecture. The potential for energy efficiency in the commercial building sector and understanding of user preferences in the residential sector leads to the establishment of a performance-based system of standards for design.

A proposal for an integrated energy and environmental planning model is outlined, which deals with the related issues of technology, population, economy and attitudes. The concept of 'efficiency with sufficiency' is introduced. Finally, the energy efficiency of building materials is examined in detail to provide a review of the energy capital costs of materials from raw state to finished products.

It is our hope that by painting this picture of the global to local, and general to specific, this book will help to disseminate the information and expertise that will assist practitioners of the built environment to avert an eco-catastrophe.

The challenge facing us in global warming

Faced with any risk the prudent course of action is to assess it, seek to reduce it and insure against it. A simple risk assessment includes the areas in which we are vulnerable and the extent to which we can prepare ourselves to reduce vulnerability. We apply such thinking to everything from insuring our home from burglary, to the design of industrial safety systems. If necessary, statutory bodies insist through guidance and regulations that we take steps to reduce risks to ourselves and others. This may include the design of the physical environment at home, the office, farm or factory, the work systems to which we are exposed and the psychological consequences of our actions. This concern extends from constructed events to those in natural environments (Taylor and Lane, 1993).

We are more inclined to take action if we can see an immediate threat, but for more distant possibilities we are inclined to delay action while awaiting evidence. In the United Kingdom the Health and Safety Executive recently issued guidance on the safety of bank staff threatened by robberies. We all know that banks get robbed. We can all recognize that facing someone with a shotgun who is threatening to 'blow your head off' is likely to be seen by most people as a traumatic event. Yet few practical steps to support the staff or customers of banks were in place even relatively recently (Lane, Woolfe and Purton, 1990).

It has taken a long period of education, persuasion and finally official guidance to generate action. Here the threat is obvious, immediate and amenable to solution. We can define both areas in which we are vulnerable and the steps we can take to prepare for them. Thus,

$$\text{Risk} = \frac{\text{Vulnerability}}{\text{Preparedness}}$$

Banks, before taking action, had to be convinced of the economic as well as the ethical case and some still required the additional incentive of a statutory push.

If action was so much delayed here, how much more difficult it is to make the case for action on global warming and the built environment. The immediate demands for expanding the built environment wash over the uncertainty of when, where and how global warming will impact.

Yet, as Paul Ehrlich powerfully argues in the Foreword to this book, there is no question that if we alter the surface of the earth and add heat-trapping gases to the atmosphere, we change the climate. The speed, degree and direction of the change is open to question, but not the effect.

These uncertainties provide an excuse for inaction but the consequences of failure to act could seriously damage the health of us all and might, as Ehrlich points out, prove devastating. Faced with so large a potential threat he argues that we really must take out insurance. Taking out insurance means in practice taking all reasonable steps to slow the influx of greenhouse gases and to limit land use to sustainable options.

Environmental accountability – will we act effectively?

In order to act we must develop a sense of 'our common future' (Brundtland Report, 1987). The importance of developing that common sense was apparent in many of the papers submitted to the UN conference on the environment in 1972. It was recognized then that we had to build on the interest of the young in their future, move towards a more educated public, prepared to modify their attitudes towards the consumption of natural resources, and generate amongst professionals a willingness to address these concerns in a multi-disciplinary framework. This latter point was seen as controversial and yet has now become essential if a sustainable future is to be possible (Lane, 1972).

As Robert Samuels (Chapter 1) argues, we can no longer support the attitude that individuals have a right to pursue their own good life without a duty of care to future generations. There is now sufficient evidence to suggest that threshold points will be reached which might set in motion chain reactions leading to irreversible destruction of our planet's life-sustaining systems. We have to adopt a 'no regrets' risk management policy and make environmental accountability a way of life. It is not easy to put a value on natural resources but environmental impact assessments focused on specific projects must now address cross-sectorial and cross-national issues if we are to be able to alter the values, expectations and behaviours of individuals and groups (Laird, 1991).

Economic progress and sustainable development are not necessarily opposed, since policies based on renewable energy can be given credit for the environmental benefits they create. There is not much evidence to suggest that we will all readily forgo the opportunities for immediate gratification in order that unknown individuals a half-century hence might enjoy similar opportunities. Yet, as computer modelling of dire predictions accumulate and natural catastrophes continue to multiply, as citizens and as decision makers we may become convinced of the necessity to stabilize growth and energy usage. Samuels points out that one area open to manipulation is the built environment. Its design and use can have a significant impact on the quality of the natural environment, and most importantly this impact can be moderated by design.

Ultimately, it is the people who use their buildings in energy-efficient ways who will make the difference, but designers need to build-in the possibilities for such use. The interaction between design and use creates the possibility of an 'energy–environment equation' for the built environment. Samuels presents such an equation, and models of the type he proposes offer a real prospect for re-directing the debate. The emphasis in his model on the relationship between energy use and energy users enables designers to consider a range of issues from the physical to the psychological environment. This multi-disciplinary perspective represents a way of thinking sought in the 1972 United Nations agenda, but sorely missing from much of the debate since.

Models such as Samuels' enable us to explore the intimate relationship between social and environmental responsibility and thereby provide one of the tools to act effectively. As such tools for understanding the relationship between the built environment, global warming and our responsibility for it become available the question becomes focused on our will to act. If we can audit our environment and specify courses of action then a declaration that we will act on the results of such audits is crucial. Leaders of the seven major industrial nations meeting in July 1989 did make such a declaration, as Gerald Vinten (Chapter 2) points out. He also maintains that major companies are beginning to appreciate their role and admit to their responsibilities in the localities where they operate.

The reasons for increasing corporate involvement in the environmental debate vary from corporate self-interest to social responsibility. The increasing costs of compliance emerging from national and international conventions and a realization that green economics may benefit the company's 'bottom line' undoubtedly plays a part. Corporate responsibility is now emerging as a major issue as company directors face fines and the threat of disqualification, i.e. debarred from practising their profession. Major disasters, such as that of the *Exxon Valdez*, with huge environmental consequences, and evidence in several recent transportation and industrial disaster inquiries that companies ignored warnings has led to calls for protection for employees who 'blow the whistle' on bad practice. But whistle-blowing when things go wrong does not represent an alternative to proactive policies and the pressure for more corporate responsibility is now strong, despite disappointing legal judgements in some cases (Vinten and Lane, 1992).

Increasingly, groups of companies are coming together to agree common sets of environmental principles to which they can subscribe. Vinten is able to present details of several such initiatives. The will to act appears to be increasing but the extent of the activity to support that will is still well short of the level needed for sustainable development. The combination of company self-interest, increased sense of social responsibility and the pressure from legislation and litigation does, at least, provide the motivation to act.

The increasing use of environmental auditing provides an additional tool. Vinten outlines the principles of an environmental audit and the steps that must be taken to ensure that it is comprehensive. The aim is to produce a management system that will provide information on environmental performance against pre-determined targets, ensuring that those targets are met. The emphasis on performance targets is a feature common to several of the contributions to this book.

However, targets do require a political will to act. That will often depends on pressure groups or democratic action to demonstrate that the agenda has changed. Will those concerned with environmental issues as citizens or members of professional bodies be able to have their voices heard? It is clear that the position varies greatly from country to country and moment to moment. Within Europe recent changes introduced by the Maastricht Treaty will make it more difficult for such voices to be heard. While the concept of the citizen's Europe is strengthened in the environmental areas, the 'scope of Maastricht and its multiplicity of procedures will make it all the harder for citizens to make their voice heard . . .' (The Maastricht Treaty and Voluntary Associations, 1993).

Professional associations will also find it more difficult to convince the European Commission to put forward proposals for action. It will only be by joining forces in a network that effective lobbying and action will be possible. There was much lobbying by environmental organizations during the discussions leading to the Maastricht Treaty to ensure that it was 'greened'. Some progress was made and the requirement to integrate environmental concerns into the implementation of community policies in a range of areas was welcomed. However, an opt-out provision enables countries to avoid action by reference to the fact that policy must take account of the diversity of situations in different regions. A step forward and a half step back.

The failure of the will to act, particularly where economic factors are believed to contradict environmental concerns, was evident in a recent survey of companies working in the field of logistics and distribution. Given their role in waste, packaging and transportation they have a crucial part to play. In a survey of the members of the Institute of Logistics and Distribution Management (1993), it was discovered that the majority wanted to respond positively to environmental concerns but felt constrained by the business climate. Over two thirds assumed that operating costs would increase as a result of environmental policies yet, where action was taken, it had a significant element of cost reduction. There is uncertainty about the benefits of action and most companies say that they will not act unless forced by legislation. Legislation is seen as having the positive effect of equalizing the competitive pressure. We all have to do it and therefore no one loses.

This combination of views is reflected in other key industries impacting on the built environment. Yet the fact that many environmentally sound policies also make good business sense is ignored. Auditing provides a way into this cycle by making the links between different aspects of the business. Environmental management systems for business are helped by systems which encourage new environmental standards (for example, the British Environmental Quality Standard BS7750).

The audit is a powerful evaluative tool; however, it cannot assist us in determining the validity of the objectives. It does not help us to understand the extent to which we really will need to embrace sustainable development. The recognition that economic factors need to reflect environmental concerns is apparent in the Maastricht Treaty, reflecting the increasing view that setting limits on pollution will not be enough to bring about sustainable development. Environmental policy will in future be considered relevant in European legislation and will cover all fiscal matters, town and country planning, and major choices in energy sources – unless the opt-out right is evoked!

The objective: sustainable development

The recognition, albeit partial, that the objective should be a move to sustainable development focuses attention on three key issues: renewable energy; urban design and transportation; and sustainable development.

Renewable energy

The first problem we face when trying to develop a policy on renewable energy is the highly distorted way that most official energy statistics overlook biomass fuels (wood and dung) and the use of techniques such as daylighting of buildings. As John Page points out (Chapter 3), this statistical failure and the consequent impact on policy formation was recognized by the UN Statistical Office (1988), and moves towards improved assessments of areas such as solar energy within national economies are now underway.

Unfortunately, public confidence in alternative energy sources has been dented by exaggerated claims but at last real and significant achievement has been made (Commission of the European Communities Programme of Solar Energy, 1989). There has to be, according to John Page, a willingness to deal in the truth about renewable energy policies, even if that means that some dreams are shattered. His agenda is clear. We must focus our attention on sustainable development, address the damage that we do through waste and create a built environment which does not subjugate all other biological species. Our approach to energy utilization offers a solution to that agenda.

Page's case, in part, is that we can integrate our approach to the two forms of energy consumption; internal, i.e. food, and external, i.e. consumption. Food production has increased at the cost of a very considerable energy subsidy. Coordination of agricultural, energy and environmental policies must become a priority. Present subsidies and costing techniques, which hide the real costs of food production and the burning of fossil fuels, create a biased market in which renewable energy sources (both biomass and solar) cannot compete. It takes the prospect of irreversible damage to the world environment to force a rethinking of energy policy. The detrimental impact of current energy policies is carefully documented by Page and he argues that it is now clear that we must have a chemical policy for the environment. Chemical planning of the built environment is almost non-existent. Proposals for changing the bias in the market are presented based on the importance of careful costing.

An important characteristic of the built environment is that the surface area involved is large. 'Solar energy can be captured at the point of demand by elements that are needed for enclosure anyway.' Yet at present poor design features mean that buildings and their supporting industries must be considered as the primary cause of potential global warming.

This case is developed by Ian Lowe (Chapter 4), who draws our attention to the idea that the future is a social construct. There are a range of models of the future and we make the choices between them. That choice is based on social values. It is a value judgement that we should try to meet our present needs without compromising the ability of future generations to meet their needs. The Inter-Governmental Panel on Climate Change (IPCC, 1990) had specified likely scenarios. Based on current policies temperature will rise at a rate greater than that experienced at any point in the last 10 000 years. Sea levels will rise, extreme events will increase in frequency.

Lowe provides a range of possible options, gives examples and makes the simple but telling point that anyone who designs buildings now presumable expects them to be around in 40 years' time. Within such a time frame climatic changes will take place, and therefore, designs need to take those into account. We need to be aware of the effect of design and orientation on solar gain for a building (principles understood for 2500 years). Professionals in the built environment cannot be excused responsibility for bad design in these areas. Lowe links building design, urban planning and transportation systems as keys to sustainable development.

Urban design and transportation – the urban village

The link between urban design and transportation systems as keys to sustainable development has not featured very strongly in the greenhouse debate according to Peter Newman (Chapter 5). The design, placement and grouping of buildings, and the consequent transportation patterns required make a major impact on the production of greenhouse gases. In a detailed survey conducted over a decade in 31 cities world-wide he is able to generate ideas on how urban design can modify transportation and contribute to the reduction of greenhouse gases. Newman's approach has influenced designers internationally and has been linked to a move towards densification – concentrating on building those urban forms with the lowest energy usage. The detailed analysis that Newman provides illustrates the folly of attempting to reduce greenhouse gases by technological tinkering. It will take an integrated design and transportation solution.

> Urban villages are the key urban design component which will enable the re-building of cities with reduced automobile dependence.

This is not, as some might argue, a measure to force people out of their cars but an approach which produces an attractive lifestyle option. He suggests that cities adopting such a route to planning would make major savings and they will gain economic as well as environmental benefits. The linking of the economic and environmental cases may provide the basis for the dialogue we need.

Development of the built environment – sustainable suburbia

Allan Rodger (Chapter 6) referring to Newman's work acknowledges it as one of the core strategies for sustainable development. Rodger presents a different approach based not on the modification of urban design but on the creation of sustainable suburbia. He too sees sustainable development as a challenge with significant implications for the future of our planet. Built environment professionals, he argues, have been slow to respond to the fundamental changes that are occurring in our environment. The way our professions are structured contributes to this in that only narrow concerns are addressed. The relevant contribution in the new situation is inherently multi-disciplinary. It also requires an understanding of systems in use, based on involvement of the user in the design process. For Rodger the built environment is not the hapless victim of climatic change but is the principal villain promoting it.

Allan Rodger urges us towards a low impact, low movement society and the modification of existing systems towards that objective. The surprising answer he proposes is the redesign of the low density suburb as an environment full of opportunity for change. His conceptual model for a productive landscape is challenging on first sight. With reflection the advantages emerge. It is certain the propositions for complexity and multi-use of suburbia will inevitably meet with resistance from professional sources and local inhabitants of low density suburbs, but his strategy has an important role to play.

The broad perspective on the interrelationship between global warming and the built environment generates a number of options at the policy level. There are also a range of specific areas of action open to us. The most promising of these are explored.

The options generated by the proposals discussed below need to be interpreted within the framework of principles discussed above. Single options are not advocated as the 'answer'. It is an interactive position that is being advocated.

Climate, comfort and culture

Patrick O'Sullivan (Chapter 7) dismisses the notion that there is some simple causal relationship between energy efficiency and architectural shape. This notion rests on the assumption that once this relationship is understood, it will enable the designer to produce streams of energy efficient buildings. O'Sullivan views energy as a commodity that building occupants use to enable them to improve the quality of their everyday lives. The use of energy, therefore, is determined by what people want to do in their buildings rather than their shape. For him the question is: how and for what reasons do people use energy in pursuing their lives in ordinary buildings? By understanding this we can find out if there is anything important about this usage that might lead to some general rules.

O'Sullivan identifies two main uses of energy. These are to produce a healthy and comfortable environment in which to live (including thermal, visual, acoustic and air quality), and to help provide the other necessities of life, i.e. water, cooking, cleaning, washing and so on – the 'power' for labour saving devices. He points out that the quantity of energy needed to accomplish these tasks is predominantly culturally determined, although there exists a strong climatic bias. The energy needed for life-support systems is dependent on two factors. These are culture – the conditions under which we are prepared to live, and climate – which determines the role of buildings in our lives.

More importantly, however, is the fact that most people live in buildings for extended periods of time (in Northern Europe 90% of time is spent in them). As a result design theories have been developed to help relate energy usage to the way individuals spend their lives. The two most important theories are climatic modification and the 'lack of discomfort' theory.

In his chapter O'Sullivan powerfully illustrates how these theories allow the design of the fabric of the building to modify the external/internal climate relationship in an energy efficient way in order to produce a healthy, comfortable and cost effective environment. In addition he postulates general rules which emerge from the application of such theories to design. He considers the impact of the 'lack of discomfort theory' on design, in particular to heat and light requirements inside buildings. Climate oriented design is then appraised in terms of 'climatically rejective design' where materials and form insulate the inside of the building from the outside elements. He goes on to discuss 'climatically interactive design' where energy efficiency is attained via 'solar displaced heating and lighting' and intermediate zones between inside and outside are generated.

Proactive architecture

From the notion of energy as a commodity embodied in architectural form in terms of climate, comfort and culture the focus now shifts to the role of the architect/designer in the search for ecological responsibility.

A disturbing but eminently relevant mandatc was given by the World Biennial of Architecture in Sofia, July 1991:

> Our planet is deteriorating. There is a global need to save the ecological base on which all life ultimately depends. We have to achieve an organic coexistence between the natural and built environment, thereby ensuring a better life for present and future generations.

Jeffery Cook, in a compelling discussion (Chapter 8), spells out the role of architects in the process of achieving an ecologically balanced lifestyle on the planet. He points out that architecture needs to be proactive – if architects do not make the case for architecture in environmental terms for the 1990s no one else will. Architecture that is benign does not go far enough. An architecture that is only neutral will not be able to heal the wounds of a global environment which is dramatically deteriorating because of accelerated abuses. A benign architecture will not restore the equilibrium of the global natural system. Cook asserts that what is required is a radical redefinition of what constitutes success in the built environment. Not just extreme change, not only a new, original approach but also a re-examination of fundamentals and a return to roots.

This assumes a rediscovery of origins. Sustainability must be the key to this radical reform. He advocates bioclimatic design – which goes beyond passive cooling and heating – as one of the stepping stones. He cites Palladio, who devoted considerable concern to bioclimatic issues and to health as the basis of design decisions. In his work, siting, orientation, building form, openings and porches have clear reference to local climate. These become elements of the composition reflecting the vision of an orderly and productive world – aesthetics enforced by performance. The classical symmetry is visible and often copied – the underlying bioclimatic order is largely invisible and hardly noticed.

Cook points to this radical role for architects through a review of the development of passive solar architecture in the USA and a discussion of the bioclimatic architecture of the 1980s. He proceeds through a chilling overview of energy expenditure, the role of the environmental industry and architectural responsibility in the 1990s. This includes the introduction of architectural service standards relating to indoor climate, air quality and so forth. Changing architectural paradigms such as ecopolis and eco-architecture, alliances with nature, healthy buildings, energy data bases for buildings and the 'rediscovery' of mud architecture are elaborated upon in the process of his argument. The Rocky Mountain Institute building in Colorado, the NMB Bank, Amsterdam, and the Permaculture Institute of Europe at Steyeberg are discussed as exemplars of ecologically sound constructions.

Commercial sector energy use

The focus shifts again in the search for energy efficiency. This time to one of the largest, yet underestimated sources of energy consumption in the built environment – the commercial building sector.

In his chapter on energy efficiency and the non-residential building sector Deo Prasad (Chapter 9) focuses on the commercial sector energy use and the consequent impact on global warming. The non-residential building sector in most countries constitutes a significant proportion of the built environment. The buildings which comprise this sector – central business districts (CBD), suburban centres, institutional buildings and the industrial and warehousing complexes – are a major cause of concern in terms of the visual environment as well as the impact on energy requirements for their construction and operation. Operational consumption of energy in such buildings is shown to be in the order of 50% for HVAC systems and 25% for lighting systems.

Consequently, the impact of these buildings on greenhouse gas emissions is significant. A considerable proportion of these requirements can be addressed by energy efficient design strategies. Prasad stresses that addressing the issue of energy consumption in terms of the total ecological construction makes good sense, not only in terms of a reduction of greenhouse gas emissions but also in a commercial sense. He demonstrates that buildings designed for low energy use – incorporated into the actual design and fabric – reduce operating costs significantly.

[These strategies include envelope/facade/window treatment, energy management systems and auditing (good house-keeping), daylight integration and use of efficient fluorescent lighting, demand-side management and building energy standards.]

This allows released funding to be placed elsewhere for commercial advantage. However, the gains do not end here. The construction of buildings, which because of their environmentally friendly design are healthy for the occupants, facilitate the kind of environment which encourages productivity and performance.

A 'performance based' approach to compliance with standards is discussed – target achievement by whatever means are considered appropriate in the local environment.

Designing for sustainability

From consideration of the effects of commercial building construction and operation on the environment the discussion focuses on the question of energy consumption in the residential sector.

In their discussion on the principles of energy efficient residential design, John Ballinger and Deborah Cassell (Chapter 10) make the point that 'all dwelling occupants' have the opportunity to contribute to reducing the greenhouse effect through the use and creation of housing. This is something that can involve individuals at the 'grass roots' level in actively pursuing the goals of energy efficiency. Ballinger and Cassell suggest that, because of this and the various characteristics that distinguish housing from other types of building, strategies which promote energy efficiency in the residential sector need to be very different for those which are suitable for other building types.

They point out that in less developed societies housing consumes little energy and is generally responsive to the environment. The form and fabric of the housing itself is used to provide thermal comfort. In most developed countries this art has been lost. Thermal comfort is achieved through the provision of energy consuming heating and cooling – mainly because of the capital costs involved in the fabric construction. Ballinger and Cassell maintain that if we are to take control of our energy consumption the model provided in the less developed societies has to be regained. However, as the less developed countries become urbanized, the danger is that the aspirations of potential occupants to live in housing which is patterned on that of developed countries becomes more pressing. It his hoped that a substantial movement towards this will be mitigated by the awareness of the importance of energy efficiency on behalf of the developing countries.

In terms of energy consumption, the development of prototypes and the establishment of regulations and standards will play an important role – which in turn will influence the context and form of housing. Ballinger and Cassell see the key to the design of housing which uses as little energy as possible and provides reasonable levels of thermal comfort as both climate and user satisfaction. The climatically appropriate house is one which responds to seasonal demands, modifying the negative effects of climate and taking advantage of the positive effects to enhance the comfort of the occupants.

A range of energy efficient solutions for the residential sector is examined. The design principles elaborated upon range from the siting of the buildings in terms of solar access (micro-climatic exposure) and the integration of building and landscaping, to building design issues relating to heat gain and heat loss. The interactivity of building systems is emphasized (orientation, glass and shading, for instance). Prescriptive 'rules of thumb' are rejected in favour of performance based solutions and recognition of the relationship between users, amenity and energy. They make the case for applied research which will enable design decisions to be made taking into account the many variables at work.

Ballinger and Cassell point out that for a design to achieve the aim of energy efficiency it is first necessary to know when and why energy is consumed in a house. It is not enough to know that more than one third of domestic energy consumption is used to meet space heating and cooling needs. Users' comfort preferences and occupancy patterns must be identified before specific heating and cooling requirements can be identified.

Home Energy Rating Schemes are proposed as a way to encourage the implementation of energy efficiency in housing. Also important are the energy conservation regulations/standards which are expressed by the establishment of mandatory requirements. When this is in place the implementation of specific measures is assured. These will vary from country to country, and climate to climate.

Of central importance in the achievement of lower energy consumption is the users. The theoretical amount of energy required to maintain a specified comfort level may bear no relationship to actual consumption. This is generally related to the users' comfort needs and desires and their willingness to be responsible for the most efficient means of operating their home. Ballinger and Cassell recommend an approach which demonstrates to people the close association between amenity and energy efficiency. Most people respond well to sunny houses with naturally lit rooms without realizing that these are the same qualities inherent in the energy efficient home.

Technology is not enough

In a discussion that comprehensively assesses technological options and sustainable energy welfare, Jorgen Norgard and Bente Christensen (Chapter 11) point out that 5 billion people live in unsustainable societies. In the near future these societies will be subjected to acute changes caused either by environmental problems or – preferably – brought about by considered action. In terms of this the industrial countries will have to reconsider their objectives for development in technology, population, economic growth and attitudes in general. All these will have to fit within an environmentally sustainable framework.

Norgard and Christensen's chapter suggests a definition of sustainability, describes ways of addressing these issues and also illustrates how energy and environmental development are closely related to other aspects of society. They consider energy as an essential area for planning, as it has a vital role in the general development of technology, economy and the environment. Energy planning for sustainability cannot be effectively carried out if it is not related to goals in economics, population, education and so forth.

In terms of this, Norgard and Christensen propose an integrated energy and environmental planning model based on iterative scenario building – i.e. a model which evolves from one scenario to another, towards sustainability. They propose that integrated energy and environmental planning will be essential in future policies as environmental constraints on development become apparent. The industrialized world's lifestyle cannot be sustained for much longer – dramatic changes will be required. Norgard and Christensen maintain that in order to achieve and maintain quality of life ('high welfare') within the constraints of sustainability, it will be necessary to restructure the planning process. The focus will have to be on end-use demand and efficiency rather than on the supply of more energy. They elaborate upon the development of efficient refrigerators, washing machines and other domestic appliances in terms of end-use efficiency. They suggest that the technological potential for reducing energy demand is great and must be explored. However, according to Norgard and Christensen, technology itself will not be sufficient to achieve sustainability and they introduce the notion of 'efficiency with sufficiency' – that is, efficient technology and appliance design, albeit necessary and desirable, is not sufficient in itself to solve the problem. Other factors in society which influence energy demand and cause environmental problems will have to be addressed and corrected. These other factors can be described as growth in population and in material standards of living (which encompasses living) – this includes economic growth and people's attitudes to material comfort.

Norgard and Christensen point out that population is mainly growing in developing countries. However, it is recognized as a problem and action is being taken to attempt to solve it. The problem of economic growth is a different matter. It is only now that this is beginning to be recognized as a significant problem by a few politicians and economists. Norgard and Christensen propose a steady state economy – where higher gross domestic product (GDP) is not the goal, but rather a better quality of life with an economy which suits the needs of the society. Examples are provided of encouraging official steps being taken towards environmental sustainability by the Danish government.

Impact of Materials

George Baird in his illuminating chapter, on materials selection and energy efficiency (Chapter 12), draws the book to its conclusion. He sets out to describe the underlying concepts and practical methods for estimating the energy embodied in a building – its capital energy cost.

Baird describes how energy efficiency has become of increasing concern to those who design, operate and use buildings. This concern has focused almost exclusively on the energy required to operate the building. These operating energy costs are readily apparent and easily understood. However, much less attention has been paid to the energy embodied in the materials which the buildings are composed of and the implications of their selection for energy efficiency. In addition he points to the environmental impact of the selection of different building materials. Until recently this has been virtually ignored because their effects tend to be indirect and in many cases remote from the building site. In reviewing the, albeit limited, research in the area he points out that the proportional importance of energy embodiment is seen to grow as building energy operational costs fall as a result of efficient design.

George Baird focuses on the environmental concerns of the implications of greenhouse gases in the earth's atmosphere as a result of increased consumption of fossil fuels. He points out that this is a trend in which the building construction industry plays a significant role. The burning of fossil fuels, especially oil and coal, is a major contributor to these emissions. Most conventional building materials are fossil fuel intensive in their production and so contribute directly to greenhouse gas concentrations. These include, for example, cement, fired brick, glass, steel, aluminium and many plastics. He maintains that these practices cannot continue if we are to achieve a sustainable future for the planet.

In his chapter Baird goes further than the assessment of energy efficient building design and operation (as these are dealt with in other parts of the book). He considers capital energy requirements of buildings and his main effort concentrates on the energy implications of using different building materials and their general environmental consequences.

The concept of capital energy and the terminology, conventions and limitations of energy analysis are explored. The energy consumed by the construction industries of two countries, the United States of America and New Zealand, is examined and compared. He also provides a detailed examination of the energy requirements and their likely environmental impact of some 10 building materials, ranging from wood to glass to paint and plastic. The application of these energy coefficients to the estimation of the capital energy requirements of a standard house is illustrated using a case study, and the effects of using different materials and components are explored. The differences between different countries are also noted.

In presenting a case study of a typical house George Baird vividly illustrates how the selection of different materials can have a significant impact on the overall amounts of energy used. The capital energy cost of a house can be up to 60% more, and this is simply due to energy coefficient differences in commonly available construction materials. This is equivalent to several years' operating energy consumption.

He concludes with thoughts on how to take account of these matters at the present time and what information is needed to improve procedures. He foresees the inclusion of low-energy embodiment criteria in building standards in the near future.

Yes, we can act effectively

The built environment plays a major role in sustaining the quality of our planet. The responsibility for furthering an understanding of the multidimensional and interactive impact of the built environment clearly lies with its practitioners and educators. Impending climatic change and the global devastation anticipated represents a threat beyond the past experience of humankind. The challenge facing built environment practitioners could not be more important since other associated issues (ozone depletion, acid rain destruction, urban smog and so on) almost pale into insignificance when compared with global warming.

As designers, architects, planners, builders, transportation managers and environmental psychologists, we need to develop the interdisciplinary perspective necessary to understand the relationship between global warming and our activities as shapers of the built environment. As organizations who create and use the built environment and natural resources we need to focus our efforts on the long-term perspective. Sustainable development is possible and much of what we need to do for environmental reasons has direct economic benefits. As policy makers we have to face up to the rhetoric that now surrounds this debate, confront the difficult questions and champion the cause of sustainable policies. As citizens, we can influence policy development. By the choices we make about our own lifestyle we can contribute to constructing a future for ourselves and for the common future of our planet.

Long before the greenhouse effect was thrust upon us traditional beliefs linked our own survival to that of the earth and sustainable development was practised. We lived from the income of the earth's resources, not its capital. We are currently consuming those capital non-renewable assets and as professionals have not yet fully grasped the acceptance of a common responsibility for our planet.

We will have to become planetary citizens of the earth, connected with one another not by country, race, religion, profession or ideology, but by a common, instinctive rhythm of the heart.

Henry Miller

References

Brundtland Report (1987) Our Common Future. Report of the World Commission on Environment and Development of the United Nations, United Nations, New York.

Institute of Logistics and Distribution Management and P-E International (1993) *Going Green*, ILDM, London.

Laird, J. (1991) Environmental accounting: putting a value on natural resources. *Our Planet*, **3**, 1.

Lane, D. A. (1972) Education in Environmental Health. *Community Care*, **4**, 149–56.

Lane, D. A., Woolfe, R. and Purton, A. (1990) *Trauma and Bank Raids*, Professional Development Foundation, London.

The Maastricht Treaty and Voluntary Associations (1993) Conference Report from Euro-Citizens Action Service, Brussels.

Taylor, A. J. W. and Lane, D. A. (1993) *Psychological Aspects of Disaster*, Trentham Books and the Professional Development Foundation, Stoke-on-Trent.

Vinten, G. and Lane, D. A. (1992) Whistle-blowing, disaster prevention and corporate responsibility, in *Psychological Aspects of Disasters* (eds A. J. W. Taylor and D. A. Lane), Trentham Books and the Professional Development Foundation, Stoke-on-Trent.

Environmental accountability: users, buildings and energy 1

Robert Samuels

1.1 Introduction

The global environmental malaise confronting us today is symptomatic of a human malaise: an attitude that individuals have a right to pursue the good life with no concomitant duty of care to ensure that future generations also partake of that good life. Whether or not an individual decides to moderate their consumption of resources, and take responsibility for the continuing quality of life, becomes an act of conscience, a quality notoriously deficient in the make-up of the human mind.

Sufficient empirical evidence of this global environmental malaise has, however, been accumulated which presents very powerful arguments for undertaking immediate precautionary measures as a minimum requirement. There is, moreover, a strong likelihood – given the observable behaviour of countless other lifeforms – that threshold points will be reached and breached in the planet's ability to adapt. Chain reactions leading to the irreversible destruction of life-sustaining planetary systems can no longer be dismissed as the ravings of environmental activists. It is the international scientific and intellectual community which is now at the forefront of the 'green movement'. This book bears witness to the commitment of leading built-environment experts to identifying the impact of the built environment on environmental quality, and, most

Global Warming and the Built Environment. Edited by Robert Samuels and Deo K. Prasad. Published in 1994 by E & FN Spon, London. ISBN 0 419 19210 7.

importantly, to suggesting solutions based on their accumulated experiences and expertise.

1.2 Conflicts of interest

Detractors, inevitably, exist. There are claims that the evidence for greenhouse warming accumulated by the IPCC and other scientific bodies is being manipulated by them for ulterior motives. Industrialists, power utilities, and other vested interest groups assert that the evidence is too uncertain (and/or still within the range of normal climatic variation) to justify a commitment to greenhouse gas emission reduction targets and global or per capita 'carbon budget' allocations (and possible income and job losses). Assertions that economic disaster will result from meeting emission targets fail not only to account for benefits accruing from a vigorous renewable energy industrial sector, but also to admit to the inherent uncertainty in economic modelling. Assumptions about population growth, per capita energy use and consumer attitudes to energy and the environment, and predictions of energy prices are all suspect – witness the extreme volatility of oil prices over the past two decades, not to mention the relative indifference of consumers to energy price changes. Maintenance of comfort, convenience and lifestyle standards appear to dominate user behaviour. There is uncertainty in climatic modelling, but uncertainty is also a major feature of economic and budget forecasts – and of political eventualities, scientific hypotheses, medical diagnoses and sub-atomic behaviour, or our comprehension of the origin of the universe! Nothing is certain. We have no choice but to act within such 'fortuitous constraints', and trust that our assessments prove to be as ethical, accurate and justifiable as possible.

Added to this is the conflict of interest which separates the developed from the developing countries, each accusing each other of culpability (over-consumption vs. over-population). There is now a suggestion that carbon budgets should be allocated to nations on a basis of their historical per capita carbon emissions or 'natural debt', i.e. an apportionment half of which is based on population and half on gross national product. For such a carbon budget to succeed, hidden carbon subsidies need to be taken into account. A wide range of manufactured goods and materials are seemingly unrelated to carbon dioxide emissions but are implicitly dependent on fossil fuel combustion. [Methane (CH_4) released by coalmining represents a second-order carbon subsidy.] Included are chemicals, fertilizers and pesticides, the vast quantities of reinforced concrete required to build hydro-electric dams and nuclear power stations, the energy expended to obtain enriched uranium (U^{235}) and, of course, the manufacture of thermal insulation materials (which play a role in energy efficiency) and of solar thermal parabolic dishes, or photovoltaic cells.

It can, further, be anticipated that those responsible for the major proportion of planetary pollution will attempt to trade-off, rather than limit, their emissions – via tradable levies with countries which are net carbon sinks, i.e. rich in forests and oceans – to avoid putting their own lifestyles at risk.

Ultimately, the argument that the cost of prevention is too high – given the relatively uncertain scientific basis upon which preventative action would have to be taken – can and must be countered with the argument that the costs which might have to be borne as a result of inaction are likely to be unsustainable. Continuing with a 'business-as-usual' scenario rather than adopting a 'no-regrets' risk management policy is to be avoided at all costs. Chaos theory tells us that unpredictability is real and that minor changes in initial states can lead to outcomes that are totally unanticipated – Lorenz's butterfly flapping its wings and creating a hurricane. We cannot afford the luxury of waiting for the most equitable and economically efficient response to be devised.

1.3 Environmental accountability

None the less, environmental accountability is a way of life, it seems, whose time has not yet come. Accountability is a strategy which is capable of gradually altering the values, expectations and behaviours of individuals and groups. If the costs to the environment are deducted from the benefits and profits resulting from its exploitation, as a matter of course, a real value for GNP, or individual project, can be attained. The extreme difficulty of putting a value on natural resources (Laird, 1991), such as clean air or water, or calculating the eco-loss sustained as a result of the extinction of a species or a rain forest, or the acidification of a lake, explains why environmental accounting seems so unattainable in practice. Environmental Impact Assessments have been around for decades, but these relate to individual projects, and are projections of possible scenarios within delimited parameters. Environmental accounting needs to deal with cross-sectorial and even cross-national interactional issues such as losses attributable to air pollution as a result of urbanization and vehicle use. Respiratory diseases and lead poisoning might ultimately be reckoned as costs of built environment development, but the lag between the emissions and the effects can be several decades, and pollution generated in a coastal city might be blown inland to downwind suburban areas – the western suburbs of Sydney being a case in point. Ironically, a growing medical service sector mobilized to cope with the consequences of urban air pollution might have the effect of increasing overall GNP growth as presently calculated.

Similarly, sulphur dioxide and nitrous oxide emissions generated in British coal-burning power plants have acidified the forests and lakes of Scandinavia. And recently a huge tropospheric ozone cloud has been detected over central Brazil and parts of Africa, apparently caused by the burning of forests in Amazonia and of savanna grasslands in Africa. Indeed, while the northern hemisphere suffers the effects of acid rain, there are high levels of tropospheric ozone, as a result of biomass burning, over much of the southern hemisphere during parts of the year. Low level ozone is not only a greenhouse gas but is corrosive and can damage the tissue of plants and humans, and accelerate building corrosion.

Proving the source of a pollutant is notoriously difficult; and it is much like comparing chalk and cheese – short-term benefits and long-term costs are far removed from one another not only in time but in concept.

On the other hand, development that is sustainable, and the use of solar and renewable energy, should be given credit for the corresponding lack of disbenefits incurred. At local level, this could manifest as subsidies and tax rebates; at the global level, as a natural capital dividend in the GNP calculation.

Classical economics, however, defines productivity narrowly and equates gains with economic progress. Ill effects of such progress are concealed by labelling them 'externalities', and because they are difficult to measure, conveniently excluding them from the calculations. Consequently, much economic growth may be an illusion because of the failure to account for reduction in natural capital and damage to eco-systems.

The attitude that humankind is master of the planet is everywhere enshrined, and the next five billion people who will walk this earth in the next half-century will not be content to accept twentieth century 'third world' living standards. They too will expect to consume at the rate that post-industrial technocratic western nations consume. Currently 20% of the world's population consumes 80% of its resources. A five-to sevenfold increase in consumption of energy and goods will be needed to raise the consumption level in the developing world to that enjoyed in the developed world (Brundtland Report, 1987). Irrespective of the extent of energy efficiency measures introduced, a population twice the size and with sophisticated expectations will exert stresses on planetary eco-systems which are unimaginable today.

Human nature has not shown itself to be particularly amenable to foregoing opportunities for immediate and personal gratification in order that anonymous individuals who might live a half-century hence might enjoy similar opportunities. The belief that technological advances will assure the continual discovery and exploitation of natural resources, or that the soil can continue to be doctored and plants genetically engineered to provide ever increasing quantities of food, must surely be a mind-set that will prove, in time, to be spurious. Notwithstanding, there are no limits to growth built-in to national economic agendas, and annual GNP growth is a priority in all nation states irrespective of their politico-economic ideology. And even proactive sustainability ideologies assume that human use of the environment inevitably degrades it, and that the goal is to limit this destruction. Improving the quality of the environment does not seem to be recognized as a practical reality – albeit ethically desirable – or it is likely to be rejected as utopian, if not irreligious.

At the same time, as advances in computer modelling improve predictability (by simulating the role of oceans, ice masses, etc., more adequately) and/or if natural catastrophes continue to multiply, it can be expected that more and more decision-makers will be convinced of the necessity to stabilize growth and energy use, or even to target reductions in the next century. One area open to manipulation is the built environment, since its design and use have a significant impact on the quality of the natural environment and, most importantly, this impact can be moderated by intelligent and benign design without reductions in lifestyle expectations and standards. The aptitudes and attitudes of the users of the built environment have a significant impact on their habitual and unselfconscious behaviours, or 'environmental roles' in those environments. The goal of this book is environmental education – to provide pointers along the way to accountability and sustainability by providing models and strategies that are achievable and believable. Intergenerational equity and planetary stewardship cannot be attained unless and until the relationship between users and the built environment is renegotiated.

A judicious mix of legislation, education and persuasion can still ensure that the planet survives us. Dr Seuss chronicled the rape of the environment in a book written for six-year olds, some 20 years ago (Figure 1.1). The 'Lorax' spoke for the forest, but the determination of the industrialist to grow rich converting the leaves of the trees into 'thneeds' eventually destroyed the forest for everyone, industrialist included. The message is the same today: no economy without environment. And the Lorax leaves behind a poignant note, engraved in stone, a single word: 'Unless'. Unless we care, unless we act, all is lost.

We can, at the very least, ensure that environmental education and environmental restoration are part and parcel of every child's and student's conceptual framework.

1.4 Social and environmental responsibility

Social and environmental responsibility in terms of the built environment encompasses two major themes of benign design: solar efficient design (SED) and ecologically sustainable design (ESD). The fundamental principle underlying both is **interactivity** – the mutual interdependence of

Figure 1.1 The Lorax. (Source: Seuss (1971) *The Lorax*, Collins, London. Published with the kind permission of Theodor S. Geisel and Audrey S. Geisel.)

all environmental systems; and the fundamental activator is energy – the product (almost exclusively) of burning fossil or fissile fuels.

Built environment systems are accountable for: the energy utilized in a vast range of primary extractive processes and the energy embodied in manufactured building materials; the energy consumed in building construction processes; the energy consumed in the day-to-day operation of buildings (heating, cooling, lighting and servicing); the energy expended in building maintenance and, ultimately, during the demolition of the building. On the one hand, the reusability/recyclability of building materials is a plus in the energy-environment interaction, while, on the other, the management of the waste produced by users of the built environment (and the hazardous wastes produced by industrial, chemical and power-producing processes) is an intractable issue of gargantuan proportions.

Over and above the issue of the energy consumed to air-condition buildings, the refrigerant devices themselves are CFC-driven, and air-conditioning thus has a dual potential to exacerbate global warming. Paradoxically, in a warmer world, comfort requirements will presumably be met by increasing the use of air-conditioners. Similarly, albeit indirectly, a considerable proportion of furniture, both domestic and commercial, is foam-blown, a procedure again reliant on the use of CFCs.

The role of forests in the maintenance of global climatic stability is fundamental, not only because they are carbon sinks but also because tropical forests are vital in the global evaporative cycle which distributes heat away from the equator towards the poles. The use of wood, as biomass for fuel, as timber for construction and furnishing, as paper for a myriad of uses, and, indirectly, in the form of coal and oil, spans most sectors of industrial and traditional economies. Moreover, woodland is often cut-back to accommodate expanding cities, the effect of which is to reduce the carbon storage capacity of the earth. However, where the wood is used as timber, carbon is locked up in the built structure and is, thus, removed from the atmosphere (until bacterial decay ultimately has its way). A sustainable plantation policy which ensures that forests are planted and harvested like other crops, might come to be considered as a rational compromise, albeit inimical to the maintenance of fauna biodiversity in those areas. If coupled with the outlawing of rain forest logging, and/or a firm refusal to import rain forest timber, the rational use of plantation timber as a built environment material could be acceptable, given that wood is both a renewable energy source and a carbon sink.

The use of thermal mass as a bioclimatic or climate-appropriate building technique needs to be evaluated in the light of the considerable amount of energy embodied in the materials themselves (concrete, brick, etc.) and the operational energy saved over the life cycle of the building.

The energy and environmental costs and benefits of a super-insulated but lightweight, timber construction needs to be weighed up against the case for thermal mass (in the appropriate climate, naturally, and thermal comfort issues aside). It is feasible, however, to employ a mix of materials, in domestic construction particularly – a heavy/massive ground floor/ living area and a lightweight, well insulated and well shaded first-floor/bedroom area, for instance. Moreover, the indiscriminate use of mass in a building is not rational. A strategic location of mass is required. Locating mass on the equator-facing or sunny side of a building but not on the sunless side appears to be a rational response, but does not seem to have been exploited by designers of passive solar buildings.

This extensive list of SED and ESD issues referred to above does not even begin to exhaust the energy-environment interrelationships inherent in the built environment. Each building exists in a context of other buildings, with an urban, suburban, regional and rural infrastructure of agricultural, industrial, commercial, governmental, and residential facilities linked by a ubiquitous transportation network which permits the movement of people, goods, services and food between those buildings. Urban design, vehicle design, building design, appliance design, etc., all are intimately related.

Metropolitan design policies and public transportation provision play a fundamental role in the energy–environment interaction. The prevalent philosophy in countries suffering the consequences of unbridled post-war suburban sprawl is focused on the notion of 'mixed uses or mixed zoning'. Socio-spatial strategies to accommodate this notion include the consolidation or densification of residential facilities in inner city areas (unless such consolidation and infill in inner city areas is 'affordable', lower income families will still migrate to the urban periphery, and suburban sprawl will continue unabated), residential development in central business district (CBD) areas, similar to European city models (the New South Wales government recently (1993) announced changes to the building code aimed at encouraging residential development in Sydney's CBD, for example, reducing ceiling height constraints in high-rise residential projects, which can increase a project's floor space by 25%, thus making the development more financially viable, and attractive to developers), the decentralization of urban facilities and services to low-density suburban areas, and medium- to high-density residential developments in the vicinity of public transport intersections or stations.

Lifestyle expectations (maintaining comfort and amenity standards), demographic developments (increasing household formation despite stable or declining populations in developed nations) and life-quality ideologies (the growth of environmental literacy and ecological consciousness, concerns for indoor air quality, etc.) must also be considered in the energy–environment interaction. Whatever 'solution' is proposed and/or implemented, family dynamics will play an important role. Families might locate in a decentralized area because work is available, but mobility in the workforce is part of the modern lifestyle, and as children mature their home range expands, and their movement patterns are increasingly away from the family centre.

The education of local users of the built environment, which might result in a consequent modification of day-to-day behaviour patterns or phenomenological routines, can fundamentally influence the amount of energy consumed globally. Providing householders with a manual explaining how best to run their homes is a necessary beginning. The recognition by individuals of the salience of their environmental roles is the key to accountability.

This multi-dimensional and holistic perspective of the built environment allows us to appreciate the very great influence which it has on land, ocean, river and atmospheric ecology, as well as climate stability. The use of the built environment exacerbates climate warming and acid precipitation – since generating and expending energy generates CO_2, NO_x, SO_2 and photochemical smog. If, as predicted, increased precipitation or drought (as the case may be), rising sea-levels, frequency of storms and intensity of winds occur, accompanied by an increased prevalence of soil salinization, air acidification and material corrosion, the built environment and its users will also suffer the consequences. In other words, the built environment enters the equation as both cause and effect. Furthermore, protecting cities against flooding, for instance, or repairing damage wrought by corrosion, or excessive rain or drought (structural damage as clay soils expand or contract, for example) consumes additional fossil-fuelled energy, and materials manufactured with such energy.

1.5 Climatic interactivity principles

Climate warming and the built environment will be discussed by authors in relation to specific aspects of this holistic perspective. Here, we will attempt to briefly elucidate general principles underlying this relationship. Three interactivity principles appear to influence climate warming: synergy, threshold and feedback.

Synergy is a **multiplier** principle, where the combined effect of two separate events accelerates the occurrence of a third event and its consequences, which are quantitatively or qualitatively different from the sum of the effects of the individual events. CO_2 increasing at the same time as CH_4 is increasing has a dual effect on temperature. It is the combined temperature rise which is critical. Added to this are temperature rises induced by NO_x and CFC emissions. Because of this synchronous interactivity, all the issues need to be dealt with simultaneously. There is a tendency amongst policy-makers – inevitably strapped for cash – to rationalize priorities in terms of opportunity costs and marginal gains, but at the risk of losing sight of the comprehensive picture. Currently, for instance, there is a strong focus on curbing CO_2 emissions, while the energy-intensive, forest-destructive and methane-producing ruminant industry expands exponentially and western hamburger-eating proclivities are imported into even such bastions of tradition as China.

Threshold is an **accumulator** principle. Adaptations to stresses in a system are frequently pseudo-adaptations, i.e. coping in the real sense of the word. Inevitably the strain of coping with individually minor stresses accumulates, albeit subconsciously or unwittingly, until the cumulative impact becomes collectively significant and a given threshold point is reached, and breached, after which the entire system can precipitously fail. The ozone depletion in the stratosphere provides us with a perfect example: CFCs accumulate over several decades, and suddenly, when a critical mass is reached, a 'hole' appears. Similarly, the warning signs of climate change might be disguised by natural occurrences such as floods and droughts produced by El Nino oscillations; yet it is difficult to dismiss the occurrence in 1992 of the worst floods in the history of Pakistan, the worst hurricane in the history of Hawaii, or the warmest decade (1980s) on record as being tell-tale events on the path to a threshold point.

The Gaian philosophy – that the earth is a superorganism operating as a single homeostatic system and capable of maintaining itself in a form suitable for life, despite catastrophic events and irrespective of the assaults on it – is based on a dynamic equilibrium view of systems which does not accord with evidence of human homeostatic systems. Rene Dubos, Hans Selye and others (Samuels, 1978) have shown that systems do not necessarily return to their previous state once stressed, that these strains accumulate, albeit over long periods of time, that some systems are more homeostatic than others, and that fatigue and ultimately exhaustion often occur. In any event, is it acceptable that humanity, animal and plant life as we know it might well become extinct, but that energy cannot be destroyed only transformed and, ultimately, the planet will recover and new life forms appear again?

Finally, feedback is the **reactive** principle. Feedback can act either as a catalyst, triggering a reaction which increases the greenhouse effect (a positive feedback), or as an event which counteracts the tendency (a negative feedback). Implicated here are the interactions of land, oceans, ice masses, clouds, plant and microbial communities with temperature change. Examples of positive feedback are: an increase in temperature resulting in increased plant respiration, in turn resulting in increased CO_2 emissions, which lead to an increase in temperature; or an increase in temperature resulting in ocean phytoplankton – a carbon sink – absorbing less CO_2, resulting in an increase in temperature. Ultraviolet radiation, increased as a result of ozone depletion, also destroys plankton, with similar positive feedback consequences. An increase in temperature that led to a melting of the tundra and permafrost zones, would, similarly, allow the carbon in the soil to become active (increasing CO_2 emissions), while simultaneously allowing soil bacteria to activate (thus generating CH_4 emissions).

Current meteorological simulation research, undertaken in Britain, and current measurements by the World Meteorological Organization (Gribben, 1992) indicate a further positive interactive process at work. Greenhouse warming in the atmosphere due to enhanced CO_2 concentrations is acting to further destroy stratospheric ozone levels because the warming of the lower atmosphere robs the stratosphere of warmth, and cold temperature is one of the key factors in springtime ozone depletion.

Negative feedback is a process that can diminish or even nullify a greenhouse effect, or can act as a camouflage event, giving the spurious impression that all is well, while, in reality, a system can be seriously out of kilter. Clouds are the classic example of a negative feedback mechanism. An increase in temperature might lead to a higher rate of evaporation, which in turn creates more cloud cover, with a consequent lowering of temperature. This is still, however, an area of great uncertainty. Cirrus clouds seem to have a warming effect (absorbing earth heat) while stratocumulus tend to have a cooling effect (scattering sun back to space).

Most importantly, the camouflage effects of acid rain, ozone depletion and volcanic activity have recently been uncovered. Aerosol sulphates which cause acidification of rain and snow which decimates lakes and forests also reflect sun back to space, thus hiding the (daytime) warming which would otherwise be manifest. The global warming which has been evident over most of the southern hemisphere of the planet in the past decade has been slow or absent over North America and Western Europe – the very areas which emit most sulphur dioxide. This anthropogenic dilemma is irresolvable. If SO_2 is reduced (scrubbed from power station stacks, for example) the result will be an intensification of climate warming.

CFC accumulation in the lower atmosphere acts as a greenhouse warming gas, while ozone depletion in the stratosphere acts to cool the earth, again masking the warming reality. Finally, volcanic eruptions belch dust and sulphuric acid droplets into the atmosphere, which reflect the sun away. The Pinatubo eruption was expected to artificially cool the earth by 0.5°C in 1992 which did indeed cool by 0.3°C to 0.4°C following the eruption. This was particularly noticeable in the interiors of continental landmasses compared with other regions (Pearce, 1993). The planetary cooling which occurred in the 1950s and 1960s can be associated with the eruptions of Mt Vesuvius, Mt Hekla, Mt Spurr, Kamchatka, and Mt Agung during that period. It is unfortunate in the extreme that these masking events have given credence to the arguments of the detractors: that evidence of greenhouse warming is absent, or that cooling has even occurred (in some places and at some times). It is unfortunate because synergy and the threshold effect are at work unremittingly, and response strategies which would otherwise be rapidly implemented are delayed or rejected as a consequence.

1.6 Environmental strategies

Response strategies are concerned with increasing environmental education and the enhancement of environmental literacy and consciousness, undertaking precautionary measures irrespective of the apparent uncertainty in the data (a 'no-regrets' strategy), undertaking preventative or mitigating measures (stabilization and reduction of emissions), and preparing for the worst, i.e. undertaking adaptive measures in order to reduce the chaos which would be borne by the ultimate victims of climate change (eco-refugees from low-lying and island countries, for instance).

It is also possible to undertake interventionist measures, such as transferring appropriate and ecologically sound technology to developing countries (rather than dumping inefficient and outmoded technologies on them); increasing family planning measures and 'attitude-altering' workshops in populous nations; introducing energy and environmental levies (carbon taxes or tradable levies and 'polluter pays' principles); establishing pricing strategies ('user pays' or beneficiaries bearing proportional costs); insisting on an environmental accounting system that subtracts environmental costs from GNP figures; pricing fuel and electricity realistically (including their environmental costs); demanding life cycle costing of the built environment (cradle to grave costing); introducing international and multi-lateral targets and timetables (and watchdog organizations to police compliance) to reduce energy production and consumption; and increasing energy efficiency in every sphere of activity.

The power utilities are also cleaning up their act by removing sulphur, carbon and nitrogen wherever possible, and are switching to gas-fired turbines (an inefficient use of a premium fuel since only half the thermal value of the gas is converted into power), but are yet to adopt and integrate renewable energy sources such as photovoltaics, wind, water and solar thermal power in their supply scenarios. Fortunately, certain enlightened utilities are adopting least-cost strategies such as fuel substitution, and demand-side strategies based on the notion that it is cheaper to save energy than to produce it. This 'end-use' objective is achieved by encouraging customers to be efficient by offering tariff rebates for energy efficient buildings, reducing rates for small rather than large consumers, supplying subsidized or free compact fluorescent lamps, and community information programmes.

Ultimately it is the people who use the energy who will make the difference. It is critical that individuals implement an environmentally apt lifestyle and adopt ecologically sustainable principles, now being taught in schools and universities, in their professional and domestic environmental roles. Built-environment designers and practitioners can contribute to this potential by ensuring that the possibilities to attain it are built-in to their designs, i.e. that buildings are ecologically sound, energy efficient and climate appropriate, that CBDs in cities include residential zones and that suburban residential zones have work and service opportunities integrated into them, that public transportation is ubiquitous and safe and efficient; and, ultimately, that accountability is an ethic which every built-environment designer and student internalizes and makes an integral part of their conceptions. Perhaps their children will have this ethic as a preconception, as an unselfconscious construct which seems as natural to them as growth and profitability are to present generations.

1.7 Efficiency and quality

In our drive to rationalize energy use and be environmentally responsible there is always the danger that this mission will blind us to other aspects of existence – the quality of life in particular. People did not take to the idea of voluntary simplicity in the 70s in the wake of the oil crisis because living standards were threatened. Altruism is forgone when lifestyle is challenged. Similarly, beseeching consumers to buy and run energy efficient houses can fall on deaf ears – saving energy is not a priority in most people's list of motivations when looking for a house to buy.

Recent Australian research carried out by the author and colleagues (Ballinger *et al.*, 1991) showed that only if energy use, thermal comfort and lifestyle amenity (natural daylight penetration, for instance) are grouped together as motivators can they account for a similar percentage of total motivations as neighbourhood and spatial motivators. Moreover, the finding that some houses designated energy efficient could use more energy than standard houses, reinforces the notion that the household as well as the house should be the focus of attention in promoting energy efficiency, and that comfort and amenity issues might outweigh energy efficiency as a goal. It would seem to be more realistic, at a microscale, to recognize energy efficient design of buildings as a potential which can be built-in to climate appropriate housing, but that associated life quality potentials are likely to be more influential in the adoption of such housing in the market-place.

Energy efficiency, as a motivator, should be clearly linked to its associated role in preserving environmental quality (environmental responsibility); and the lifestyle qualities of SED houses should be emphasized, i.e. being warm in winter and cool in summer, the delight of abundant indoor natural light, having close indoor/outdoor contacts, etc. Such an approach is deemed more likely to influence consumer housing choices, and economies of energy can be anticipated as a consequence.

Another, even more recent, research project carried out by the author (Samuels and Ballinger, 1992) examined the comparative psycho-social well-being of office workers exposed to either very energy efficient fluorescent lamps of 'good' spectral quality, or to relatively inefficient but full spectrum, daylight-simulating lamps of excellent spectral quality. Respondents worked for nine months (unwittingly) under the different lamp types. At the end of the period, those working under the full spectrum lighting had significantly less experiences of headaches and fatigue at work (*inter alia*). Certain sick building syndromes and seasonal affective disorder syndromes (i.e. consequences of natural light deprivation) seem, thus, to be influenced by the quality of artificial lighting. Here, again, the issue is one of opportunity costs, i.e. benefits achieved vs. benefits forgone – quality and social responsibility (health, performance) having to be traded off against energy efficiency and environmental/ecological responsibility.

1.8 Towards an energy–environment equation

During the past decade, the author's research and teaching interests have been inter-disciplinary in nature and scope, but have lacked a strong central focus. The 'greenhouse effect', however, has provided such a focus, and the critical mass around which a considerable proportion of the various threads can now be woven. The result is an 'energy–environment equation' (EEE) (Figure 1.2), offered here as an evolving synthesis, and multi-dimensional evaluative model, of the built environment. It is not an equation in the mathematical sense, but a set of interrelationships. In terms of this model, Eco-quality and Equity (or socio-quality) in the built environment are functions of Energy Use, i.e. the situational consumption of energy (which fuels and emissions, in what quantities; which functions served, in what physical forms; what hidden agendas?, i.e. the temporal-spatial and politico-economic context) qualified by the characteristics, ideologies and behaviours of the Energy Users of the built environment (the existential and socio-cultural context, grouped under the energy user category).

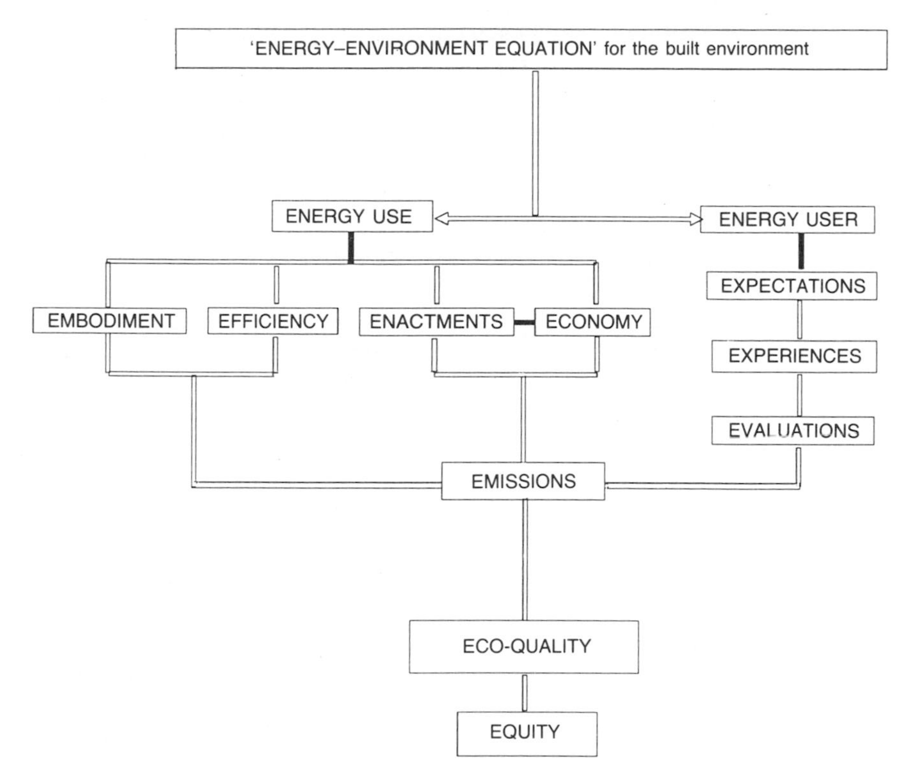

Figure 1.2 The energy–environment equation for the built environment.

Eco-quality refers to the cumulative impact of each of the facets on: climate stability; atmospheric, air, water, ocean and soil quality; deforestation and desertification; and the preservation of natural habitat and the biodiversity of animal species. Ultimately, the quality of the global eco-system is a function of the nature and quantity of the emissions and effluents which are attributable to the interactions between energy use and energy users.

Equity refers to the egalitarian distribution of human health, well-being and satisfaction; equality of opportunity and responsibility regarding energy use; and intergenerational sustainability. Such issues are inextricably related to the quality and health of the global eco-system.

Some equity issues which might be included as non-quantitative modifiers of the energy–environment model are: social isolation and lack of facilities in suburbia, which might result in an extended use of private vehicles as people seek stimulation and services elsewhere; air pollution, noise annoyance, and anomie in inner cities, which residents periodically seek to escape by travelling to the country or coast; and the social severance effect of railways on neighbourhoods and communities, often overlooked when public transit systems are proposed as a solution (light railway systems, e.g. trams, are not effective over long distances). It would also be remiss, when proposing land-use mixing and densification in suburban areas as an energy–environment solution, not to consider equity issues such as residents' concerns about loss of privacy (being overlooked), or perceived threats to the low density residential character associated with the 'quarter-acre dream'. At a national and global scale, the rights of less developed regions or countries to approach equality with developed areas by increasing their use of energy would also have to be considered within the equity factor.

Embodiment refers to the energy consumed during: the primary extractive processes, the manufacture, and the transportation of building materials; and, ultimately, the demolition and disposal of buildings and other physical infrastructure. The energy embodied in the manufacture of trucks, on-site equipment such as cranes and pile-drivers, and earth-moving machinery must also be accounted for as part of the capital energy consumed in the built environment. Similarly, all vehicles and transit systems (including ships and aircraft) used to service a metropolis incorporate a hidden energy expenditure. Even a move to solar powered vehicles would not reduce this expenditure. Capital energy use is also a function of the durability of materials and structures, and their recyclability potential.

Efficiency refers to 'operational' energy consumed as a consequence of: metropolitan design (urban and suburban configurations or settings – hence land-use zoning, siting, density, integrated land-use/transportation planning, etc.); solar efficient design, i.e. the climate-appropriateness of buildings in terms of the control of heat gain (insolation) and heat loss (insulation); *in situ* construction processes and site preparation operations; and the day-to-day energy expended in the running of the built environment (heating, cooling and lighting, appliances and equipment, vertical transportation in high-rise buildings, industrial machinery, and transportation – the movement of people and goods).

Enactments refer to energy policy and performance criteria, such as solar access site-design standards; building, appliance and materials efficiency guidelines; home energy rating schemes; (reduction) targets, treaties and conventions set as goals (such as the 14 priority areas elaborated in Agenda 21 at the Rio Earth Summit, June, 1992) and the biodiversity and climate change conventions signed at Rio (Boyd, 1992) and, paradoxically, licences to pollute. Such enactments sometimes have the force of law, or are prerequisites for building and development applications, but usually (particularly in the international arena) they are proposed as ideals towards which to strive. Discouraging the importation of rain forest timber for building purposes, rather than prohibiting it outright, would be a case in point.

The major element which influences the uptake of such enactments is economic imperatives. The environment vs. employment debate pits the environment lobby against industrial vested interests and government at national level. At international level, developing nations refuse to have the 'north' or the 'west', as the case may be, regulate the use of their natural resources (frequently over-exploited in order to service a crippling foreign debt). On the other hand, developed countries refuse to constrain their international competitivity by taking polluter-and-user-pay policies on board.

A further aspect of economy to take into consideration, although difficult to quantify, should be the global environmental cost of a development, over and above the localized cost/benefit analysis which can be provided by an Environmental Impact Statement. Hidden subsidies for fringe suburban developments in the form of free infrastructure services of individual plots (estimated at $40–70,000 in Sydney), is now being recognized as an area to address in a better global costing system. A proper price for petrol would also go a long way to reducing the major impact of private vehicles on the quality of the environment.

The interaction of the energy-use elements is inevitably modified by the impact of the energy users; and the sheer number of users is by no means the only consideration.

User expectations refer to aptitudes and attitudes concerning energy and the environment. Aptitudes reflect the degree of environmental education, ecological consciousness and energy literacy which a population possesses, and are therefore potentially able to bring to bear on their actions in the environment. Attitudes refer to : cognitions, values and goals; standard of living anticipations; motivations, preferences and priorities; beliefs about the salience of individual acts in the global scenario, and consequent behavioural intentions; and a sense of responsibility/accountability (which might be embedded in an individual's *Lebensweld* or life-world ideology, and/or be encoded in cultural norms and ethical systems with which individuals are expected to comply).

Experiences are the acting out, in specific settings, of environmental roles, for example, socio-spatial and temporal-spatial behaviours such as seasonal-specific behavioural patterns relating to the 'running of a house'; or the undertaking of monitoring activities relating to what, where, when and how energy is consumed in a given building or setting. These are cognitive or selfconscious activities. As important are the phenomenological or *un*selfconscious habits and routines or world of everyday experiences and activities of all individual consumers of energy in all built environments.

Evaluations, in this context, refer to : perceptions of thermal comfort (in buildings and outdoor spaces) and satisfaction with amenity or lifestyle quality (daylight penetration, for instance), both of which influence energy use. Some aspects of satisfaction with performance in the workplace (related to artificial lighting and air-conditioning, in particular) are also energy related issues. To this must be added the degree of acceptability of the physiological health and psychological well-being consequences of energy–environment interactions – an equity issue.

Assessments of the appropriateness of alternative energy–environment interactions are the ultimate evaluations which users of the built environment can make. The quality of their energy–environment education and the quality and extent of the information provided to them will, thus, play a fundamental role in the sustainability of global ecological systems in the 21st century.

Finally, Emissions (exhausts and effluents) are related to embodiment and efficiency, enactments and economy, and expectations, experiences and evaluations, and refer to the indirect and hazardous consequences of energy generation and consumption. In particular: greenhouse gases and acid precipitation as by-products of the burning of fossil fuels to generate electricity, photochemical smog from vehicle and industrial emissions in urban areas, ocean oil spills, ozone depletion from the ubiquitous use of chlorine-based chemicals (e.g. CFCs used for refrigeration and air-conditioning, and in the manufacture of PVC plastic products – which also produces organochlorine waste), and ionizing radioactivity related to the use of uranium as a fuel. Fuel type, quantity, quality, renewability, and hidden subsidies are issues of major relevance, as are waste management policies and technologies.

On a practical level, the energy–environment equation provides a 'holistic' and interdisciplinary, although not necessarily complete, overview of the eco-quality issue. It also attempts to address the issue of social responsibility, or the effect of the built environment on the quality of life. However, while it is currently possible to determine energy embodiment and emission levels, establish policy issues influencing enactments and economic constraints, and elicit user perceptions and conceptions, the inclusion of equity issues in the model will require extensive research in order to formulate a checklist (open-ended by definition).

Ultimately, equity could become an adjunct to the EEE, whereby the eco-quality outcome might be weighted by the degree to which a socio-quality assessment is positive or negative. Some trade-off factor would be required. If, for example, epidemiological rates of respiratory health were influenced by a legislative ruling on the efficiency of vehicle motors, or the prevalence of mosquito-borne diseases increased as a result of climate warming, an indirect impact of energy use on eco-quality would be implicated – given that humans are also animals living in an eco-system, and their quality of life is contingent on the quality of that eco-system.

The implications and complications of such issues are enormous. None the less, an analysis of the interactions of as many EEE facets that are measurable at this stage, in a given situation and setting, should be able to serve as both an indicator of the status quo and a predictor of the likely direction in which that situation will evolve. Appropriate inputs at vital stages of the equation could, ideally, be catalysts for a change in direction.

Utopian as this may sound, unless we recognize that there is a problem, establish models and find techniques with which to diagnose it, and move rapidly towards creating sustainable solutions, the prognosis for the maintenance or enhancement of global living standards, particularly after 2050, when population and CO_2 concentrations are expected to have doubled, is stark.

1.9 Conclusion

Social responsibility and environmental responsibility are intimately related, and must both be recognized and incorporated in a multi-dimensional energy–environment accountability scenario that is to address the total well-being of the planet and its inhabitants.

And there is no time to lose. In a geological time-frame, humanity could be in the last split second of its existence. Within the lifetime of the generation born today the planet could either be terminally ill, or on a path to sustainable recovery. The choice, ultimately, is ours. For better or for worse, humankind is now capable of influencing the geological fate of the earth.

To rephrase Einstein, God does not play dice with the universe . . . but humans do!

Acknowledgement

Acknowledgement is made to Fred Pearce for his illuminating articles over the years in *New Scientist*, which have helped educate this author.

References

Ballinger, J. A., Samuels, R., Coldicutt, S., *et al.* (1991) A National Evaluation of Energy Efficient Houses. Final Report to Energy Research and Development Corporation, Canberra. Project No. 1274, Solarch, UNSW.

Boyd, D. (1992) UNEP after Rio. *Our Planet* **4** (4), 8–11.

Brundtland Report (1987) Our Common Future. Report of the World Commission on Environment and Development of the United Nations, United Nations, New York.

Gribben, J. (1992) Arctic ozone threatened by global warming. *New Scientist*, Nov. 28, 16.

Laird, J. (1991) Environmental accounting: putting a value on natural resources. *Our Planet*, **3**, 16–18.

Pearce, F. (1993) Pinatubo points to vulnerable climate. *New Scientist*, June 19, 7.

Samuels, R. (1978) The psychology of stress: impact of the urban environment. PhD thesis, University of Reading, UK.

Samuels, R. and Ballinger, J. A. (1992) Quality and Efficiency in Lighting – Social and Environmental Responsibility. Final Report to Pacific Power, NSW, Solarch, UNSW.

Environmental auditing and the built environment

2

Gerald Vinten

2.1 Introduction

An audit can provide no more valuable service than ensuring the survival of the planet. This may seem to claim too much for an activity which is much more modest in scope. There will be increasing numbers of false messiahs as we approach the year 2000, and environmental auditing can do without being amongst them. Despite the need for caution, an audit has a valid role to play, and it can bring about improvement in the built environment.

2.2 The natural organization

When the leaders of the seven major industrial nations met in July 1989 they all agreed in a declaration that industry has a crucial role to play in preventing pollution at source, in waste minimization, in energy conservation and in the design and marketing of cost-effective, clean technologies. Respectable companies are beginning to appreciate that with their use of fundamental natural resources, they must increasingly become actively involved in conservation and sustainable development. They will also admit to responsibilities in the localities in which they make their money, employ their staff and use public services.

Global Warming and the Built Environment. Edited by Robert Samuels and Deo K. Prasad.
Published in 1994 by E & FN Spon, London. ISBN 0 419 19210 7.

The reasons for corporate involvement may be stated as:

1. Commercial self-interest. Urban and rural decay will result in less spending by residents, wasted land development opportunities, problems of recruitment, and the constraining factor of unpredictable shortages of basic materials. Companies can no longer take for granted the automatic existence of a natural infrastructure, yet without this their whole endeavour may be jeopardized.
2. Company image. This applies particularly to companies with a prominent public profile. According to US research an integration of corporate responsibility programmes into an overall public relations campaign can lead to an enhancement of status in the eyes of the public, increased produce demand and help with recruitment. In an interview with IBM the importance to existing employees of environmental concern and action by their employer was stressed. This view was based on regular Employee Attitude Surveys, which are very much part of IBM's open communication system. Shell believes that building contacts with environmentalists helps the company's image and modifies hostility to the company when accidents occur.
3. The extent of regulation. Responsible companies wish to act voluntarily in advance of regulation. Thus du Pont aimed to cut hazardous waste it produced by up to 70% over a two year period. These companies realize that they must secure and retain general public consent for their activities, and so they do not oppose regulation. They believe that it helps their cause if they can demonstrate that their activities are rigorously monitored by an independent body. They can also witness the competitive disadvantage in which it places cowboy industries.
4. The use of their industry-specific knowledge to reduce environmental uncertainty. Thus IBM has as part of its environmental policy that where skills and resources are relevant, it will assist government to develop solutions to pressing environmental problems. For example, computer modelling, and technologies such as remote-sensing image processing and geo-referencing are key to the study of climatic change and ozone depletion. Similarly the chemical industry is privy to vital information, since environmental damage can be broken down into a series of chemical effects. The Industry Council for Electrical and Electronic Equipment Recycling (ICER) was launched by 13 top businesses, including IBM, ICI, ICL, Thorn EMI, British Telecom, Hewlett Packard and Boots. It resulted from a report in December 1991 about the disposal and recycling of electronics discards. It aims to find ecologically acceptable and cheap ways of dealing with obsolete electronic equipment. Estimates are that within the square mile of the City of London alone 2000 tonnes of obsolete

computers and other office equipment are thrown out each year. The Corporation of London is to join in the scheme, and is the first local authority to do so, but others will quickly follow this lead.

5. The need to identify environmental costs and expenditure. If the recommendations of the 1989 Pearce Report in the UK are adopted, companies would be obliged to practise more sophisticated costing systems, and to pay the full cost of environmental damage. There will also be a need to undertake cost-benefit analysis as to the impact on the bottom line of profitability. Costs can be saved by forestalling expensive, time-consuming conflicts with environmentalists, planners and inspectors.
6. The necessity to keep pace with changing societal values. Membership of environmental bodies is increasing dramatically, and some such organizations now have larger paid-up memberships than the political parties. Friends of the Earth membership actually doubled in 1988–1989. The activities of Greenpeace feature regularly on the news and in the media. The 1992 Greenpeace Environmental Survey showed a degree of public concern greater than that reflected in the political agenda. As the expectations gap is necessarily closed, companies may find themselves subject to ever tighter regulation and standards. Although the survey did not constitute a representative sample, it is indicative of the trend of informed opinion formers. Ranked in order of descending importance were the issues commonly in the news which gave rise to the greatest concern:
   (a) destruction of the ozone layer;
   (b) river and sea pollution;
   (c) pollution from cars;
   (d) the greenhouse effect;
   (e) transport and disposal of radioactive waste;
   (f) whaling;
   (g) transport and disposal of toxic waste;
   (h) disappearance and killing of dolphins.

   Although they were not offered in the survey, respondents also mentioned rain forests, noise pollution and energy conservation.
7. The advertising agencies now have special and growing sections devoted to 'greenism'.
8. Government itself is beginning to take a green stance, and this will influence the corporate sector. In May 1990 the British Schools Minister announced that environmental issues are high on the political agenda both in this country and abroad, and that environmental education in schools should be concerned with all aspects, from the local to the global. At the same time the Department of Health issued a press release on green, but not recycled, paper, announcing a new

campaign to befriend the environment 'Towards a Greener National Health Service'. This involved:

(a) protection of the environment – safer clinical waste disposal methods;
(b) conservation of energy;
(c) improvements to the physical environment – £10 million for the interior and exterior of buildings;
(d) use of lead-free petrol.

9. There is evidence that investment that is environmentally beneficial disproportionately and positively impacts on a company's share valuation, and that the converse is true of companies that degrade the environment.
10. Company directors will increasingly face the threat of fines and disqualification.
11. There are corresponding liabilities for the companies themselves. Companies with a presence in the United States need to be aware of the Federal Sentencing Guidelines which became finalized on 1 November 1991. There is an appreciable British presence there. The Guidelines provide judges with a formula for sentencing organizations for various types of white-collar crime. The Guidelines, which are applicable to all organizations, mean that a range of crimes from violations of employment laws, mail and wire fraud, and commercial bribery to money laundering and environmental offences, will attract fines and sanctions that can amount to hundreds of millions of dollars. These sums may seem extreme, but even conservative estimates of the long-term harm some organizations inflict on the environment, would suggest that such sums are in fact an under-estimate rather than an over-estimate of the sometimes irreversible harm to the planet. The Guidelines do encourage sound commercial practice by allowing for reduced sentences for organizations that can display evidence of an 'effective program to prevent and detect violations of law'.

2.3 The Valdez principles and other initiatives

These principles were assembled by the Coalition for Environmentally Responsible Economies on behalf of the Social Investment Forum of Boston in Massachusetts, following the 1989 *Exxon Valdez* disaster, and launched in the UK by the Green Alliance in November 1989. The Forum is a group of 325 socially concerned bankers, brokers, analysts, environmentalists and others, with combined assets of $350 billion, wishing to prevent future environmental disasters, and giving companies a common set of environmental principles to subscribe to. The 10 principles are:

1. Protection of the biosphere.
   We will minimize and strive to eliminate the release of any pollutant that may cause environmental damage to air, water, or earth or its inhabitants. We will safeguard habitats in rivers, lakes, wetlands, coastal zones and oceans and will minimize contributing to global warming, depletion of the ozone layer, acid rain or smog.
2. Sustainable use of natural resources.
   We will make sustainable use of renewable resources, such as water, soils and forests. We will conserve nonrenewable natural resources through efficient use and careful planning. We will protect wildlife habitat, open spaces and wilderness, while protecting biodiversity.
3. Reduction and disposal of waste
   We will minimize the creation of waste, especially hazardous waste, and wherever possible recycle materials. We will dispose of wastes through safe and responsible methods.
4. Wise use of energy.
   We will make every effort to make use of environmentally safe and sustainable energy sources to meet our needs. We will invest in improved energy efficiency and conservation in our operations. We will maximize the energy efficiency of products we produce or sell.
5. Risk reduction.
   We will minimize the environmental, health and safety risks to our employees and the communities in which we operate by employing safe technologies and operating procedures and by being constantly prepared for emergencies.
6. Marketing of safe products and services
   We will sell products or services that minimize adverse environmental impacts and that are safe as consumers commonly use them. We will inform consumers of the environmental impacts of our products or services.
7. Damage compensation.
   We will take responsibility for any harm we cause to the environment by making every effort to fully restore the environment and to compensate those persons who are adversely affected.
8. Disclosure.
   We will disclose to our employees and to the public incidents relating to our operations that cause environmental harm or pose health or safety hazards. We will disclose potential environmental, health or safety hazards posed by our operations, and we will not take any action against employees who report any condition that creates a danger to the environment or poses health and safety hazards.
9. Environmental directors and managers.
   At least one member of the Board of Directors will be a person qualified to represent environmental interests. We will commit man-

agement resources to implement these Principles, including the funding of an office of vice-president for environmental affairs or an equivalent executive position, reporting directly to the CEO, to monitor and report upon our implementation efforts.

10. Assessment and annual audit.
    We will conduct and make public an annual self-evaluation of our progress in implementing these Principles and in complying with all applicable laws and regulations throughout our worldwide operations. We will work towards the timely creation of independent environmental audit procedures which we will complete annually and make available to the public.

A useful channel of communication, established in 1991, is the Advisory Committee on Business and the Environment (ACBE). This consists of 25 senior executives from UK based companies, and is chaired by John Collins, chairman and chief executive of Shell UK. Its purpose is to define an environmental action plan for business and to promote closer dialogue with government on environmental policy. The government responded to ACBE's first report in November 1991. The report contained recommendations in three key areas: recycling, global warming and environmental management. The target audience was ACBE members, the wider business community, and both central and local government. Recommendations in the first two areas require significant government action, such as tax changes, and HM Treasury and other departments are committed by an action plan to respond. Future agenda items will be:

- greener growth: the extent to which improved environmental performance leads to enhanced economic viability for individual companies;
- promulgation: the particular problems facing smaller companies, unable to invest in the development of environmental management;
- environmental management systems: the relationship between two new voluntary standards in environment management, the EC eco-audit proposal and the British Standards Institute standard.

The second report contained recommendations from three ACBE working groups on commercial and export opportunities, environmental management and global warming. There are also working groups on recycling and the financial sector. The ministers welcomed the positive approach the committee was taking towards reducing the use of company cars as 'perks', and the role business can play to minimize CO_2 emissions from vehicle use. An environmental Best Practice Guide was published at the end of 1992.

Another worthy initiative stems from the Confederation of British Industry, who founded the UK Environment Business Forum in January 1992. This took up the government's wish for a 'Green Club' drawn from the UK's 1000 largest companies. The Secretary of State for the Environment contacted chief executives throughout the UK seeking their commitment to a strategy of:

- environment review;
- target setting;
- public reporting.

The CBI's 'Green Club' is operated as a voluntary scheme open to all businesses. It mirrors the CBI's existing Regional Environment Committees. Forum members have at their disposal the resources of the CBI as well as collaborating institutions such as Business in the Environment, the Environment Council and the Groundwork Foundation. The Forum will nurture commitment to:

- appointing a board level director with responsibility for environmental management;
- publishing a corporate environmental policy;
- setting clear targets and objectives;
- public reporting of progress in meeting these objectives;
- ensuring adequate environmental training;
- establishing appropriate 'partnerships' to promote the objectives of the forum, especially with small companies.

The Institute of Management's Code of Conduct and Guides to Professional Management Practice of September 1991 contains, in the first, a requirement to provide on request information to any committee or sub-committee of the Institute established to investigate any alleged breach of the Code, and, in the second, two sections relating to the environment. The first of these sections, which is Section 5, concerns UK society and the environment and states that the professional manager should:

(d) Seek to conserve resources wherever possible, especially those which are non-renewable.
(e) Seek to avoid destruction of resources by pollution and have a contingency plan for limiting destruction in the event of a disaster.

Section 6 concerns overseas societies and environments, and indicates that the professional manager should:

(a) Be aware of the management implications of global environmental issues.
(b) Decline to solve UK problems of pollution and processes by exporting them unchanged to the detriment of the quality of life of other societies.

2.4 Legal perspectives on the environment

The auditor will be concerned to ensure compliance with laws and regulations on the environment. In the US context this will be legislation such as the Water Quality Improvement Act of 1970, the Noise Control Act of 1972, the Toxic Substances Control Act of 1976, the Clean Air Act Amendment of 1977, and the Resource Conservation and Recovery Act of 1984. For the UK, it is in particular the 1990 Environment Protection Act. For the European Community it is article 130R of the 1987 Single European Act which states that:

1. Action by the Community relating to the environment shall have the following objectives:
   (i) to preserve, protect and improve the quality of the environment;
   (ii) to contribute towards protecting human health;
   (iii) to ensure a prudent and rational utilization of natural resources.
2. Action by the Community relating to the environment shall be based on the principles that preventive action should be taken, that environmental damage should as a priority be rectified at source, and that the polluter should pay. Environmental protection requirements shall be a component of the Community's other policies.

2.4.1 Contamination of the Land Register

Environmental work is driven by both business factors (such as the development of inner-city sites that are contaminated, for profit) and by legislation (the forced clean-up of polluted sites). The recession of the 1990s has reduced the number of business-driven projects, but equally it has led to the demise of traditional industries, and increased the amount and cheapness of potentially contaminated land available for development. Because of the scarcity of good building land, and the decline of long established traditional industries (such as steel and shipbuilding), commercial, industrial and residential developments are often sited on contaminated land. These sites have to be investigated and treated such that the contamination does not present a hazard to the user or present an ongoing pollution risk like groundwater contamination. In other cases operating industrial sites and waste tips must be subject to remedial treatment. Landfill sites in particular were in the past frequently not engineered, and many were not only causing groundwater pollution, but also giving rise to methane gas risk. With established off-the-shelf design techniques and solutions for new landfills, it pays to design properly in the first place. It is in the design of remedial works for contaminated sites that the problems and the costs arise.

It has been estimated that there may be 75 000–100 000 contaminated sites in the UK, with a clean-up bill of up to £30 billion. Under section 143 of the 1990 Environment Protection Act, there is a requirement for local authorities to maintain public registers of potentially contaminated land. In a departure from the original intention of a comprehensive list, David Maclean, the Environment Minister, reduced in 1992 the listing requirement to about 20% of that first envisaged. To be included now is only land seriously contaminated by heavy industrial processes and needing remedial action. A small number of use criteria specify the production of asbestos, coke and gas, lead and steel, and oil refining as coming under the category. Thousands of lesser cases – scrapyards, farm buildings, dungheaps, wood workshops – will not now be included. The register, introduced in 1993, is two-tiered:

- land that may have been contaminated;
- land subject to cleaning operations.

It is possible to move from the first category to the second, but not to leave the register altogether.

2.4.2 The Litter Code for local authorities

Much of the work of local authorities in removing litter and cleaning roads is subject to both the Local Government Act 1988 and the Environment Protection Act (EPA) 1990. The 1990 Act set higher standards to be adhered to. Section 89 requires local authorities and other specific landowners as far as practicable to keep all land to which the public has access and which is open to the air clear of refuse and litter.

Legislation is beginning to bite, and proactive companies are already beginning to adjust their strategy, plans and procedures to cope. Companies such as Hoffman-La Roche, Switzerland's third largest chemical company, the Finnish conglomerate Nestlé, Volvo in Sweden, Imperial Chemical Industries in England, Opel in Germany, Philips in the Netherlands, and IBM and Johnson and Johnson in the US are just a few of a growing number of companies that see competitive advantage as well as moral justification for the green audit.

2.5 Enter the green audit

Prince Charles, heir to the British throne, speaking on British television on May 23 1990 on a programme he produced on the environment, emphasized the need for proper audits and accountability. He reinforced the message in his keynote address to the World Commission on Environment and Development (Brundtland Report, 1992; Brundtland, 1987) on April 22 1992, and in subsequent conversation with the writer.

2.5.1 The objectives of the green audit

The international Chamber of Commerce has defined environmental audit as 'a management tool comprising a systematic, documented, periodic and objective evaluation of how well environmental organization, management and equipment are performing with the aim of helping to safeguard the environment by:

- facilitating management control of environmental protection;
- assessing compliance with company policies which would include meeting regulatory requirements.'

From this we can see that the concept is wide. It is a total strategic, cradle-to-grave process, which some prefer to call 'conception to resurrection', underlining the 'design to recycling' aspect. The main aim is to produce a management system that will provide information on environmental performance against predetermined targets, ensuring that those targets are met, and that one remains ahead of the pack. The adoption of this definition by the National Environmental Auditors Registration Scheme, and by the Institute of Environmental Assessment has added extra authority to this definition.

The scope of the audit can be comprehensive, take on a range of issues, or address a single topic. Where international audits are carried out by a central team, it is likely that a range of topics will be covered to minimize costs. The audit may be restricted to simple compliance testing, or extended to a more searching and inter-professional audit. The audit will address the operating environment, health and safety management, product safety and quality, loss prevention, and minimizing resource use and adverse effect on the environment.

2.5.2 The content of the audit

The following stages have proved to be successful in practice.

1. The pre-audit.
   If this type of audit is new to the organization, then it will be worthwhile to make a quick scan across the activities contained in 5. below. This will highlight areas requiring attention, suggest how the audits will fit in with other work, what resources may be needed, and the training involved for all those who will participate, both auditors and auditees. This situation will also pertain to a multinational company wishing to expand home audits to subsidiaries abroad. If the audit is part of a continuing programme, then a review of past achievement against plans and new needs, will suggest areas for attention.
2. Setting objectives.
   (a) Define goals. From the pre-audit the major audit concerns will

have emerged. It is now necessary to determine where change is feasible, and the impact is greatest per unit of expenditure. Four basic questions should set the tone:

(i) what are we doing?
(ii) can we improve?
(iii) can we do more?
(iv) can we do it more cheaply?

The aim will be to optimize the trade-off between company efficiency and negative impact on the environment.

(b) Decide audit scope. The focus may be on the entire organization (comprehensive audit), a department (activity audit) or site-by-site (site audit). The pre-audit will have given some idea as to which of this, or mixture of these, would be worthwhile.

(c) Ascertain regulations, standards and state-of-the-art technology. Apart from present regulations and standards it is also important to consider those that may be in gestation. Professional, government, environmental and trade bodies, as well as line managers, will be sources of information.

Recent technological and product developments that can improve environmental performance also need to be accessed. The organization can then be judged against this benchmark.

(d) Priority setting. It is realistic to start with projects where either the risk and materiality are high, or the payback period is quickest. Energy conservation and tariffs are an example. The cost of water may well rise sharply, and conservation here is increasingly necessary. Such projects will establish credibility, and looking after the cents will take care of the dollars, and lead to the audit of the larger and sometimes more intractable issues.

3. Preparation.

(a) Select the audit team. It is important to form an interdisciplinary team which together can tackle the science and technology involved, as well as having a knowledge of departments implicated in the audits.

(b) Determine the need for external assistance. Where internal audit or the organization lack the expertise, a mixture of in-house staff and independent external consultants needs to be considered. Some large companies use special dedicated in-house staff only, and have environmental audit groups. Others, such as Union Carbide and Allied-Signal, whilst having their own specialist teams, include an independent consultant on most of their audits.

(c) Establish the terms of reference. This provides an authorization to proceed, as well as a notification to all employees who should then be encouraged to participate.

(d) Agree timescale and budget. This places bounds on the audit in relation to the projected benefits and the perceptions of top management, and provides a target for which the environmental audit manager should aim.

4. Draw up audit documentation.
   Data needs to be collected in a common format wherever possible, and interview and other questionnaires constructed. Computerized processing will also necessitate form design. Standardized documentation could also encourage self-audits, and become part of the normal managerial processes.
5. Areas to audit.
   These will vary from organization to organization, but below are example areas likely to be common to most organizations, and which are currently the subject of environmental audit attention. Three major headings have been devised: organization strategy; functional areas; and operational factors. Some examples from each are provided in turn.
   (a) Organization strategy.
      (i) Overall environmental policy.
         Does it exist?
         Has it the support of the board of directors, top management and workforce?
         Is it a regular agenda item for board and other meetings?
         Has a senior executive been given the responsibility to ensure implementation of the policy throughout the organization?
         Does this senior executive report directly to board level?
         Have environmental liaison officers been appointed for each major area of the business?
         Do they have the resources and authority to monitor compliance with and achieve implementation of organization-wide environmental policy?
         Are the coordination mechanisms between the liaison officers and the senior executive adequate?
         Are there means within the policy of resolving conflict between the environmental policy and other organization policy?
         Is the policy communicated with shareholders, employees, customers, suppliers, local politicians, neighbours and control authorities?
         Did it take their views into account?
         Is the policy document dated?
         Is it revised periodically and as soon as internal or external change dictate?
         Are all interested parties consulted?
         Is there a consistent ecological strategy for the organization?

Are the organization's objectives set with due regard for ecological factors?

(ii) Accident and emergency procedures.
Are there adequate contingency plans for dealing with accidents and emergencies?
Is the public relations department ready to communicate with employees, neighbours, the Press and others?
Are there controls to ensure that only PR statements that will stand up to independent scrutiny are permitted to be issued?

(b) Functional areas.

(i) Finance.
Is environmental impact taken into account in all investment decisions?
Are ethical and green investments chosen wherever feasible?
If an investment is likely to increase pollution, has it been investigated whether there is a lower-pollution alternative or whether the likely costs have been included in the project costing?
Is short-termism avoided, where a longer-term perspective will lead to higher environmental dividends?

(ii) Production.
Has the earlier replacement of existing production plant and the acquisition of new, non-polluting machinery been considered?
Are ecological materials and processes in use?
Are clean technologies, with better input–output ratios, in use?
Are useful materials and heat recovered?
Are emissions minimized by post-production environmental protection measures?
Can raw material specifications be altered without unduly affecting product quality and to improve environmental aspects?

(c) Operational factors.

(i) Discharges (including air, water and noise).
Are process controls and management systems adequate to ensure compliance with legislation and future objectives, as well as to avoid complaints?

(ii) Site tidiness
Are measures taken to eliminate litter and sources of untidiness inside buildings, outside and in the immediate surroundings?
Has suitable landscaping been considered to improve the appearance of the site?

(iii) Transport.

Are staff encouraged to use public transport and could you increase the availability of public transport by providing additional company funded services?

Is there an inventory of all means of transport used by the organization?

Is the organization using the most efficient and environmentally sound systems for transporting goods, people and materials?

Are only low-pollution vehicles purchased?

Are existing vehicles re-equipped and serviced with environmental considerations in mind?

(iv) Water use.

Is there a water management policy?

Is there a water management officer responsible for implementation?

Is water used efficiently?

Are regular reviews carried out to determine water consumption, leakage and waste water patterns?

Are there permanent monitors installed to assess usage?

Are ever tighter targets set for the future management of water?

Is there evidence of achievement and improvement?

Can you reduce consumption by using alternative cooling methods or controlling leakages more effectively?

Are there alternative supplies available that could reduce metered costs:

surface water abstraction?

use of a borehole?

local groundwater?

cheaper non-potable water from a local water utility?

Where possible, is inferior quality reused/treated water used for industrial uses such as washing, pre-rinsing or indirect cooling of a product?

Is such water non-corrosive and non-scale producing?

Is pipework strengthened where there are high pressures or temperatures?

Is pipework insulated where there are extremes of temperature?

Is soft or softened water – with its reduced heating costs – used where appropriate?

Is water stored on site as a risk-management tactic against interruption of supply?

If so are conditions such as to safeguard against contamination by micro-organisms and other sources?
Are there regular inspections for contamination, and loss through leakage of evaporation?
Is optimal quality and quantity of water for various phases of the production cycle achieved as per manufacturers' recommendations for their machinery and equipment?
Are water and energy-saving devices fitted in washrooms and toilets?
Have ways been sought to reduce water usage in washing plant, floors and vehicles?
Is recycled water used wherever possible when water is used for ornamental, decorative or horticultural purposes?
Arc hoses fitted with automatic cut-off valves?

(v) Recycling.
Are all opportunities considered?
Could redundant, used products be recycled?

(vi) Wastes.
Are steps taken to minimize, eliminate or recycle it?
Are recycling opportunities being lost by failure to segregate different types of waste?
Is waste disposed of responsibly?

(vii) Energy use.
Are electricity, steam, water and gas metered at the major points of use, and targets set to reduce their usage?
Is full use made of alternative energy sources, such as landfill gas, waste derived fuel, solar and wind energy, and combined heat and power?
Are energy conservation schemes in existence and adequate?
Are buildings and plant properly insulated?
Can savings be made in heating and lighting costs?

(viii) Occupational health and safety.
Is there an occupational health department?
Does it provide advice on practical accident and injury prevention?
Does it look after employees' psychological needs?
Does it advise on health education in home and family?
Does it advise on stress reduction and time management?

6. Data processing.
The data collected needs to be analysed, and then presented in a form easily understood and from which clear conclusions may be drawn. Possible conclusions should be discussed wherever possible with the staff directly involved.

7. Reporting.
   Significant defects, as soon as identified, need to be reported immediately to the Chief Executive Officer for quick action. The more routine findings should be presented to the Board of Directors as an 'executive summary' and with clear recommendations for action. These should include an estimate of cost, resource needs, the optimum time for introduction, and when the next review should take place. The Board should be invited to agree the recommendations and authorize implementation.
8. Implementation.
   The decisions now become part of corporate policy as an agreed Action Plan. This needs to be publicized extensively internally and employees made aware of the implications in their everyday working practices. The Action Plan will include deadlines for action.
9. Post-audit review.
   A periodic check will be made that the Board decisions have been fully implemented, with any non-compliance reported back to the Board. The opportunity will also be taken to undertake a quality assurance review of the audit process itself.

2.6 Conclusion

It is clear that the advantages of the environmental audit are immense. It is able to provide an independent check on the nature of the greenhouse effect, and make a valuable contribution to enhancing the quality of the built environment. The development of the environmental audit is inexorable, and those who neglect its potential do so at their peril.

References

Brundtland Report (1992) Report of the World Commission on Environment and Development of the United Nations, United Nations, New York.

Brundtland, B. (1987) *Our Common Future*. Open University Press, Milton Keynes.

Human use of renewable energy to improve the global environment in the post-Brundtland era 3

John Page

3.1 Introduction

> Nature possesses properties and forces whose discovery and right use appear to be among man's highest tasks, because they have the power to make his labour more fruitful.
>
> Von Thunen, *The Isolated State*, 1820

Returning home on a troop ship after the war that had killed 43 million people, thankfully alive, sitting on the deck in the sun, I read a book on physical chemistry by Getman and Daniels (1943). It explained to me for the first time scientifically the interactions between light and chemical matter. I was fascinated. I was not to meet Dr Farrington Daniels until the Conference on Solar Energy in Arizona in 1955.

3.2 The changing situation

When I heard I had got the Farrington Daniels Lecture Award, I decided to reread Daniels' book *Direct Use of the Sun's Energy* (1964). Once again, the chapter on the role of chemistry particularly interested me. Farrington Daniels' preface makes three key points. I will quote:

> Research in the field of solar energy is unique in several respects. First

Global Warming and the Built Environment. Edited by Robert Samuels and Deo K. Prasad. Published in 1994 by E & FN Spon, London. ISBN 0 419 19210 7.

> it cuts across many different sciences and branches of engineering – physics, chemistry, astronomy, chemical engineering, mechanical engineering and electrical engineering. Often an area between different fields of science that has been neglected becomes a fruitful field of research. Solar energy is an example.
> Second, it holds promise of leading rather soon to benefits for human welfare. Scientists formerly took little thought of the social and political impact of their work – but all this has been changed since they developed nuclear energy and made atomic warfare possible.
> Third, it can be carried out in small laboratories with inexpensive equipment.

(The contrast was being made with nuclear power research.)

Paradoxically the biological sciences were omitted from Dr Farrington Daniels' list of key associated disciplines, in spite of the fact that biomass utilization is the energy application that currently contributes most solar energy to the human economy. The Brundtland Report (1987) suggests that biomass and hydropower, estimated at 2 TW, provide currently about 21% of the power consumed by humans worldwide, 15% being biomass and 6% hydropower. Unfortunately most official energy statistics tend to completely overlook wood fuels, and other biomass, like dung. Nor do they value applications like the daylighting of buildings. The consequent distorted energy statistics provide a bad basis for national energy planning, especially in rural areas of developing countries. For example, Hall (1989a) states that, in 1981, the 830 million rural Chinese derived 84% of their energy from biomass, mostly agricultural residues, i.e. 0.8 tonnes of biomass/capita/year. It is easy to forget that information is the proper base of sound policy. Hall estimates that 35% of developing country energy currently comes from biomass sources. The UN Statistical Office (1988) recently organized a very worthwhile meeting in Rome to discuss the statistical problems to be overcome in achieving an improved assessment of the energy contribution of non-conventional fuels. I found myself with the interesting task of presenting a paper for discussion on the improved statistical assessment of solar energy utilization within national economies.

Unfortunately the 'rather soon' benefits have taken rather longer to consolidate than Farrington Daniels thought likely in 1964. We have basically discovered that the development of a reliable worldwide solar energy industry requires a very considerable collaborative international effort. The density of solar energy is relatively low, so solar collectors have necessarily to be area extensive. Good engineering design is therefore essential to achieve economy of materials use in collectors. Constant exposure to the natural environment tends to be a relatively stressful engineering exposure, though we are well used to considering it in my discipline of science and building. Therefore sound materials science knowledge is critical to the achievement of long life reliable performance. The various types of collectors, hot water, photovoltaic, etc., must assemble into systems that will provide predictable levels of system performance in conjunction with appropriate energy storage systems, which must also provide reliable performance. Control systems must be adequate to meet the demands put on them. It must be possible to meet maintenance requirements in economical and sensible ways. Design codes and standards for consumer protection are also needed. As the payback periods for solar energy systems are relatively long, future commercial success will depend very much on the public perception of whether or not durable and reliable system performance over long periods of years is actually likely to be provided by the products being offered in the market-place. The governmental and industrial investment needed to achieve reliable solar technologies has had to be relatively large, though small compared with more favourably funded areas like nuclear power. Unfortunately public confidence in solar energy was seriously undermined in the 1970s in many countries by the various groups and manufacturers who offered unreliable products to the public. The situation was often aggravated by totally unreasonable performance claims. The big achievement of solar energy research and development over the past 10 years, well exemplified by the Commission of the European Communities programme on solar energy (CEC, 1988), has been the establishment of a proper engineering foundation for the evolution of the vast solar energy industries that will be needed to secure the long-term future of humankind. We should not underrate that achievement, and be proud of it.

I have maintained for some time that while small may be beautiful, bigger is likely to be better in terms of reliability and performance. I see the future of solar energy as requiring an opting into the advantages of large-scale industrial processes and into international scientific and technological collaboration to secure a sustainable future, and not in opting out. Such intercountry collaboration, working towards common defined goals, has certainly proved very successful in the context of the European Community solar energy programme. There are simply too many people in the world for the anarchic process, with all its scientific and engineering deficiencies, to succeed. Individual communes may survive happily in rich countries well endowed with land, but the energy needs of the vast cities of the world simply cannot be sustained that way. Too often 'green dreams' on solar energy are quite out of touch with 'green realities'. It would be nice to be able to solve our problems easily, but it will not be like that. I would see one of the tasks of the International Solar Energy Society to better inform 'green' politics on the real issues and difficulties in implementing solar energy policies, even if that does shatter some existing dreams. Let us deal in the truth.

3.3 Waste cycles and the structuring of future energy policy

There are some key issues, which we now recognize as very important, which were not perceived by Farrington Daniels. The population pressure issue was perceived, so were the finite resource problems, and also the special needs of the developing countries. The atmospheric limitations on the expanded use of the currently adopted human waste disposal cycles, like carbon dioxide from combustion, methane from animal stock and other wastes, fluorocarbons from aerosol propellants and refrigeration plants, and the unsolved waste cycles of nuclear power plants, however, were not perceived. Farrington Daniels was an excellent ground bound chemist, who worked at a time when there were not the proper scientific opportunities to study the photochemistry of the earth's atmosphere in appropriate detail. However, even at that time, Dr Abbot and his collaborators at the Smithsonian Institute, had laid the physical foundations for the radiative study of the earth's atmosphere. The systematic study of the detailed environmental chemistry of the atmosphere and the oceans had yet to be comprehensively developed.

Now that a broader environmental awareness has developed, we have to re-evaluate the future of solar energy utilization in relation to other energy sources within the new broader global environmental context. Central must be the issue of sustainable development, which is the main theme of the Brundtland Report (1987). Economically we are now forced to examine and properly cost our waste cycles in terms of the damage they do. In addition we must ask, 'Is all development to be for humans alone?' When living space is short, our attitudes to other living species become especially important. Is there to be simply a habitat for humans, with the rest of the biological world hopefully in helpful subjugation? Are we to destroy the riches of evolution, ourselves as well? Malthusian human pressures on food resources will raise well nigh insuperably difficult problems for all other species on earth. I see it as part of the solar energy agenda.

3.4 Energy profiles of our present world

In order to decide on the best policies to pursue internationally in solar energy applications in various parts of the world, it is necessary to consider the current energy demand profiles of the world in which we live, as well as the solar radiation climates of different regions. The statistics given in the Brundtland Report (1987) will be used here. Energy consumption currently depends very much on national wealth. Human societies require energy in two forms, energy for internal consumption, i.e. food, and energy for external consumption. Food production per capita in recent years has fortunately increased in all continents save Africa. This advance has been achieved using two main energy dependent processes, namely the application of increasing amounts of fertilizer to plants bred to use higher fertilizer levels effectively, and secondly, by applying water to irrigate crops, which implies energy for pumping. The huge application of chemical fertilizers, in Europe especially, has led to considerably increased crop yields, but also to substantial nitrate pollution of underground water supplies, and over-fertilization of rivers and shallow seas close to river mouths. In 1989, algal blooms in the Adriatic produced serious disruption. Irrigation, used in non-sustainable ways, has led, in many areas, to substantial saline pollution of the ground. Nor is the situation favourable from the energy point of view. The energy subsidy in the European farming food chain is very considerable indeed. Blaxter (1986) suggests the support energy used in the UK food industry is in the ratio of 9.8 units of primary fuel expenditure to each unit energy of human food consumed. The production of animal protein is very energy intensive. Modern nutritional science is demonstrating that the contemporary high animal fat diets of most developed countries are far from healthy, and the human need for animal protein is often grossly exaggerated. As Thoreau, the pioneer of alternative rural life styles observed in 1854: 'One farmer says to me, "You cannot live on vegetable food only, for it furnishes nothing to make bones with"; and so he religiously devotes a part of his day to supplying his system with the raw material of bones; walking all the while he talks behind his oxen, which, with vegetable made bones, jerk him and his lumbering plough along in spite of every obstacle.'

It is clear that fertilized food biomass is not a net energy productive process. Therefore the main current biomass energy opportunities lie in making much more effective use of biomass food residues, like straw, and organic waste materials (for example refer to Molle, in CEC, 1988). Such a parallel policy can also help with the environmental problems that so easily arise from the failures to handle organic waste cycles effectively, for example, lifeless deoxygenated rivers. One must also stress the low net productive efficiency of biomass systems for energy production, i.e. less than 0.75% solar energy harnessed, in a world likely to become increasingly short of agricultural land per capita. Biomass production per hectare would fall subsequently if ever the fossil fuel energy subsidies implicit in the fertilizers were to be removed. There is therefore a strong conflict between continued rapid population growth, and cropped land availability per head for food supplies, unless the fossil fuel fertilizer prop can be substituted in some other way. The land shortage in the developing world is already putting strong pressures on

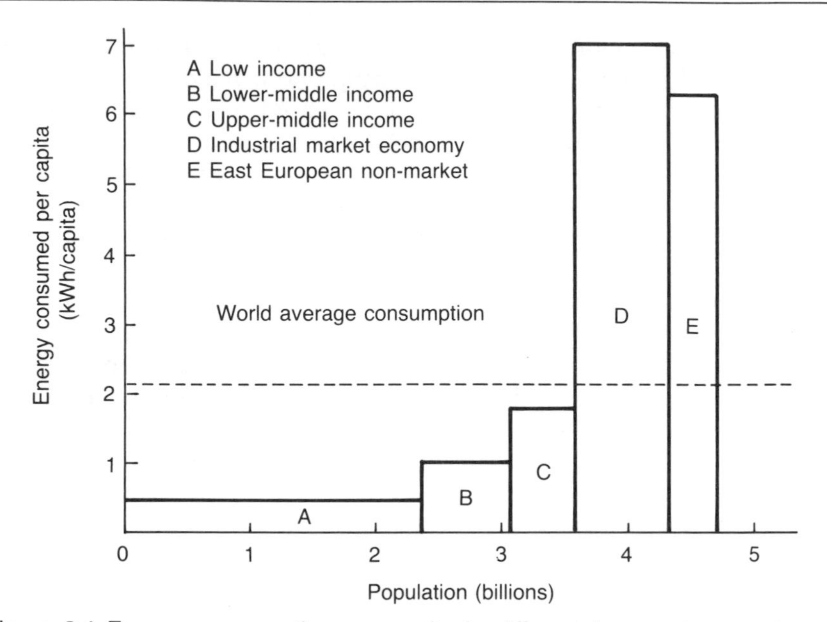

Figure 3.1 Energy consumption per capita in different income type regions in 1984 versus total population in each income type group. (Source: Brundtland Report, 1987.)

the undeveloped frontiers, i.e. the natural forests, while, in highly fertilized parts of the world, land is being taken out of production due to production surpluses. At the same time, troubles with excessive nitrogen in the water supplies remain due to the high fertilization of the continuing cropping areas. Unfertilized biomass production achieved through reforestation of the land being set aside from fertilized food overproduction could offer the two-way benefits of materials production and long-term carbon storage. Agricultural policy, energy policy and environmental policy are evidently not very well coordinated in these important areas.

Turning now to external energy, Figure 3.1 shows the mean energy consumption per head in 1984 in countries at different income levels plotted against population in each income group. The low income data is certainly distorted downwards by the failure to take proper account of external biomass consumption in national statistics. The total mean world energy consumption in 1984 was estimated at 9.94 terawatts (TW).

We now need to consider possible future energy budgets. Before the carbon dioxide crisis was perceived, high energy growth world scenarios were widespread. The Brundtland Report (1987) quotes, as an example, the 1982 35 TW world scenario for the year 2020 produced by the International Institute for Applied Systems Analysis, which was based on producing 1.6 times as much oil, 3.4 times as much natural gas, and nearly five times as much coal as in 1980. Nuclear capacity was to increase no less than 30 times the 1980 levels. The second scenario for 2030 discussed in the Brundtland Report (1987) was a lower world energy scenario for consumption of 11.2 TW, achieved through a high conservation strategy. Energy demand for developed industrialized countries was set at levels reduced from the current 7.0 TW in 1980 to 3.3 TW in 2030, with developing countries demand rising from 3.9 TW in 1980 to 7.3 TW in 2030. Such lower scenarios demand an energy efficiency revolution. In 1980 it was suggested that renewables provided 1.7 TW. The real issue for the International Solar Energy Community is whether this contribution can be significantly expanded by the new applications of solar energy to, say, 5 TW by 2030, and how precisely it might be done. I would also point out if one makes use of solar energy, there is no need to impose energy consumption quotas on atmospheric environmental grounds.

Basically we now have a situation where we know our technologies work, but using present costing techniques, which largely ignore the rapidly growing environmental costs of fossil fuel energy provision, our technologies cannot compete in a biased market, while all the time the environmental situation is deteriorating. We also have a situation, where the terrestrial environment itself is likely to change, so we also have to ask the question, 'How are energy demand and solar energy supply likely to be affected by the expected changes in the earth's atmosphere?'. Global warming will not be uniform. The greatest changes are predicted to occur in winter at higher latitudes. Temperature changes close to the equator are expected to be far smaller. The higher temperatures imply more water vapour in the sky, and a redistribution of cloud concentrations. So climatic change can impinge on solar radiation availability.

I would like now to look at solar energy utilization within its overall global environmental context. Figure 3.2 illustrates the relationships between sun and humankind in the context of the atmosphere, the biosphere, the hydrosphere and geosphere. Humankind can benefit from these interrelationships, if we proceed wisely, but can effect vast destruction if we proceed unwisely. An understanding of nature and its balances must guide solar energy applications development policy.

We need to look at pollution at three levels: pollution of the indoor environment, pollution of the local environment and pollution of the global environment. Energy policy, and both local and national environmental policy, are inextricably interlinked, both one with another and also with issues of human health policy. They are also interlinked with issues of achievable policies for sustaining the future health of the global atmosphere and of the oceans. Combustion is a significant cause of both outdoor and indoor pollution, and hence of much unnecessary human ill health. It is also a cause of acid rain. A natural product, like wood,

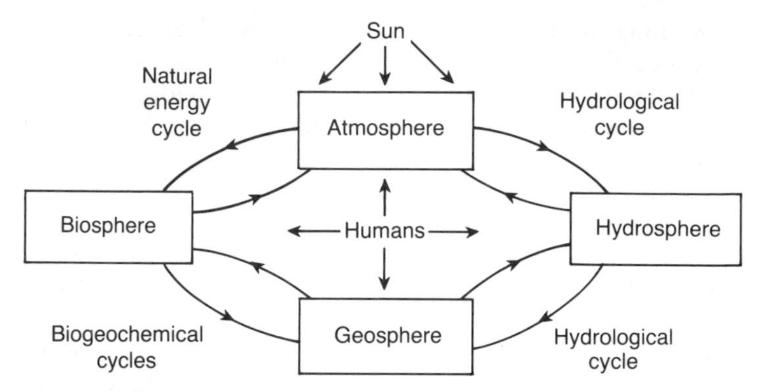

Figure 3.2 Components of the global environmental system on which the sun and humankind impact.

burned inappropriately, and much of it is burned inappropriately, is very damaging to health. Considerable respiratory ill health worldwide is linked with the use of fossil and, especially, biomass fuels inside buildings. Women in developing countries are particularly at risk from wood fuel based cooking activities (WHO, 1989). Natural gas combustion from stoves can produce significantly harmful indoor concentrations of oxides of nitrogen (WHO, 1989). 'Clean' needs to be defined. It is not synonymous with green. Clean has a biological meaning and a chemical meaning. The biological meaning has to consider the multiplicative capacity of living organisms. So, given the right environment and appropriate nourishment, a few organisms can soon produce many. Current studies are showing clean renewable energy – for example, the substitution of passive solar heating for the 'dirty' indoor combustion of fossil fuels for heating, and simultaneously providing effective ventilation to keep the indoor environment biologically and chemically clean – can make important contributions to the respiratory health of non-smoking individuals, as well as improving their thermal comfort. Simultaneously they help secure the health of the global ecosystem, in addition to contributing to the energy economy. Renewable energy applications now have to be set in this broader environmental context. For a start, such benefits to human health demand economic evaluation.

Global atmospheric issues have risen rapidly up the world agenda of priorities. The emerging crisis in the global atmospheric environment has a chemical and radiative cause. The physical effect of local pollution on the intensity of short wave solar radiation has long been clearly perceived. In an age of heavy smoke, too much atmospheric pollution was all too visible. What were not clearly perceived 25 years ago, were the interactions between the natural radiative environment and the trace atmospheric chemical contents, for example the upper atmospheric

UV-C radiative impact of the sun's radiation on human-made covalent compounds that are very stable in the lower atmosphere, principally CFCs and halons. The new gaseous threats are invisible. The implications for global warming due to the greenhouse effect must now be viewed as potentially very serious (UNEP, 1987a). Atmospheric carbon dioxide, increasing steadily, is the most important cause, contributing about half the global warming effect. In developing acceptable future fossil fuel based energy policies, we are now trapped by the adverse effects of their waste products. Here I must stress the important difference between continuously recycling to the atmosphere the current carbon inventories held in short-term biomass crops, from adding permanently to that atmospheric carbon inventory by drawing on fossil fuel reserves, or cutting down and burning the global standing forest reserves, grown over many years. The warming effect is further increased by methane, coming from cows, rice paddy fields and biomass wastes. This production increases in proportion to population. Nitrous oxide from combustion and soil denitrification adds further warming. Finally we must take note of the predicted effects of the various CFCs, which are estimated to contribute nearly one third of the warming effect. The need to control CFCs is very significant for future world air-conditioning and refrigeration policy.

Increased carbon dioxide concentrations have important biological implications (Hall, 1989b). They increase plant production, through carbon dioxide fertilization, both on the land, and, possibly, particularly importantly for the world carbon balance, in the seas. The changes should also increase the amount of carbon photosynthetically fixed in the terrestrial biosphere per unit of respired water. Solar energy utilization in marine biosystems, especially its role in removing atmospheric carbon to the deeper layers of the ocean, has received relatively little study in spite of the vast area of the oceans. This topic is of great significance for any attempts aimed at achieving a more acceptable redistribution of the global chemical inventory of carbon.

High carbon dioxide concentrations in inspired air have adverse health effects on humans, altering the acid/base balance (WHO, 1989). The effect would presumably extend to other animals. We cannot therefore afford to be complacent about zoological aspects of atmospheric carbon changes.

The other important global atmospheric issue is the increased transmission of UV-B radiation due to reductions in the amount of stratospheric ozone (UNEP, 1987b). This layer protects humankind, plants and animals from the damaging effects of UV-B radiation. At the same time we are seeing increases in tropospheric ozone due to photochemical reactions in the lower atmosphere. One particularly significant stratospheric reaction involves the photochemical release of chlorine in the stratosphere from the CFC gases, discharged at the surface, gradually migrating upwards. The breakdown products can then enter into chain reactions destroying some of the ozone. Much poorly considered and exaggerated speculation on the resultant biological effects from such changes has recently occurred. I have attempted to give a balanced assessment concerning the probable near term impacts of ozone changes on the health of humankind in my recent work (WHO, 1989).

It is now clear that a chemical policy for the environment is just as important as a physical policy for the environment. It matters a lot where our chemical inventories reside in the world, and precisely how they might become adversely redistributed by inappropriate developmental policies initiated by humans. 'Chemical planning' of human settlements is currently almost non-existent in contrast with 'physical planning'. Biological policy is interactive with both physical and chemical environmental policy. The biological world is often seriously damaged by the nature of the chemical environments that result from current fossil fuel use, including tropospheric ozone and acid rain. Huge chemical damage to human-made objects like buildings and cars also occurs. Repairing this damage not only consumes additional resources, but also requires additional energy-rich materials to do it, so further increasing the world carbon dioxide production. The atmospheric, the terrestrial and the marine systems interact, with few people giving proper attention to the needs of marine systems in developmental planning terms.

I will only make four brief comments about nuclear energy. First, the nuclear industry only supplies a very small proportion of the world's energy. Its expanded use consequently will have very little influence on the global carbon dioxide problem in the next 30 years. Secondly, the nuclear industry has not solved its own waste product problems. Thirdly, the economic decommissioning problem at the end of station life is not satisfactorily resolved. The steady accumulation of nuclear sarcophagi lying around for centuries is unlikely to prove socially acceptable. Fourthly, the economics of nuclear energy depends on substantial inputs of fossil fuel to produce the necessary construction and shielding materials. This fossil fuel subsidy adds to the carbon dioxide inventory. This does not seem to have been evaluated.

We therefore come back to a future based on renewable energy. Issues of the integration of policy between different sectors of government are always complex, but they will have to be firmly addressed to meet the goals of sustainable development outlined in the Brundtland Report (1987). It follows that energy policy is too important to leave to narrowly conceived energy departments acting in isolation. It also follows that renewable energy research and development has to broaden its terms of reference, so the wider environmental interactions are better considered. Multi-disciplinary intersectorial working is going to be essential to achieve the sustainable development goals envisaged in the Brundtland Report (1987).

3.5 Economic appraisal

The economic appraisal of conventional energy systems has been dominated by the supply side of the conventional energy industries. These industries have often received massive subsidies in terms of loans at special rates, capital write-offs and government sponsored R & D at massive levels compared with the renewables. This is particularly true of nuclear energy. Especially interesting from the point of view of solar energy applications is the US study 'The hidden costs of energy' by Heede, Morgan and Ridley (1986). In 1984, they state, the tax money used to subsidize energy producers in the USA was $523 for every household in the United States. It was the combined effect of direct agency outlays, tax breaks, loans and loan guarantees and an array of federal support programmes. More than $41 billion of the total $44 billion US subsidy was provided to mature energy technologies, which long ago reached commercialization. All the emerging renewable energy sources – solar, wind, waste to energy, ethanol, wood and geothermal – received only $1.7 billion.

Looking to the environmental side of the balance sheet, the environmental costs of conventional and nuclear energy have been largely transferred to other sectors of the economy. The true costs of different conventional energy supply systems have not been evaluated, and imposed on energy users. The polluter has not been paying. As far as the utilities are concerned, the economic costs of waste disposal have ended at the 'factory gate/chimney', regardless of the further damage to society and the world environment that resulted. This situation acts very much against the interests of the renewables, because they receive no credit for the environmental benefits they confer. The issue of social costs of new and renewable sources of energy against conventional energy has been studied in a Commission of the European Communities sponsored project by Hohmeyer (1988). Neglecting the carbon dioxide global warming issue, Hohmeyer estimated the net social costs of conventional electricity generation, as compared to wind and photovoltaic energy, in the range 0.06–0.17 DM kWh^{-1} more expensive. This compares with the market price of electricity supplied to consumers in the Federal Republic of Germany of 0.25 DM kWh^{-1}.

The concept of a carbon tax, currently under discussion in Europe, would significantly shift the economic balance between conventional energy and renewable energy, other than long-term storage biomass, i.e. forest trees used as fuel rather than as materials. A carbon tax on the use of long-standing timber for fuel would discourage deforestation. In fact there is no reason in principle why a carbon tax should not be imposed on burning-off rain forest for cattle ranching that cannot be sustained. This might encourage proper management of rain forests as productive eco-systems of considerable human benefit. A carbon tax also would significantly raise the price of nuclear power stations, because of their very large content of materials like steel and concrete, produced using carbon dioxide-producing fossil fuels. In view of their carbon consumption reduction role, carbon-bearing thermal insulation materials ought to carry a negative tax, i.e. a subsidy, to accelerate their use in conservation and carbon dioxide reduction. Looking at anaerobic carbon bearing waste cycles, for example urban and agricultural wastes, one might impose an environmental methane discharge tax on untreated wastes to encourage the proper use of methane for human purposes. The carbon dioxide generated by burning off methane would have only 5% of the greenhouse effect of unburned methane. As so much paper eventually ends up as methane generated from anaerobic processes in waste tips, perhaps there also should be a carbon tax on paper.

The economic situation will shift even further in favour of renewable energy, if power stations, in addition to extensive desulphurization, are forced towards carbon dioxide removal as well, to offset excessive global warming. The International Solar Energy Society has tended to give a low priority to systematic strategic studies of the overall economics of renewables. The environmental benefits have been talked about, but have not been quantified economically in the context of national and international energy policy. I attach considerable importance to getting more environmental economists interested in our progress. So I must add economists to Dr Farrington Daniels' list of relevant professions.

3.6 Solar energy and building

An important characteristic of buildings is that their surface areas are necessarily large, and so, by appropriate design, solar energy can be captured at the point of demand by elements that are needed for enclosure anyway. No additional separate land for energy collection is needed. Energy use in buildings and for the construction of buildings, which are normally made of energy intensive materials, fired to give long life durability, and CFCs associated with air-conditioning, probably account for at least half of the greenhouse gas emissions, possibly more. Buildings and their supporting industries therefore must be considered the primary cause of potential global warming.

The simplicity and low cost of passive heated solar buildings has made them among the most acceptable immediate solar applications together with flat plate water heaters in suitable climates. One must also note the growing interest in natural lighting, as a scientifically based design activity, belonging within the scope of solar energy research. The European Community now has a well defined programme in this area (CEC, 1990). A well designed direct gain building of heavy mass offers good daylighting without summer overheating, as well as significant energy savings in cold weather, and a healthy indoor environment with low pollution impacts. Scientifically, passive solar building research and development has been concerned to a large extent with rediscovering what ancient builders knew by pragmatic experience and providing quantitative design methods and manuals to enable designers to confidently produce designs of reliable performance. Improved techniques for energy storage, improved building surface properties and improved fenestration systems are available. Fortunately we now seem to have moved through the phase when heating alone was considered by solar enthusiasts, in favour of a more balanced all year round environmental approach. This approach places particular emphasis on the appropriate control strategies to adopt for the 'solar pass' through elements at different seasons of the year. Advances in process control are enabling the solar control systems to respond in logical ways to the solar inputs, freeing the solar building from the need for excessive user control. In the future, it should prove possible to link building controls to weather forecasts to establish short-term anticipatory control. Such measures will reduce energy demands, and consequently the outputs of greenhouse gases.

Precisely why will we need energy for building purposes in a warmer world? Figure 3.3 plots the current world population distribution against the current annual mean temperatures. A relatively small proportion of the world's population lives at cool temperatures. A large number of people already live in hot climates, and we are now predicting that climates will become even hotter. This means the current unsatisfied demand for comfort cooling must grow substantially, especially as so much of the estimated world population growth is in the hotter areas of the world. The conventional energy intensive way of dealing with overhot buildings and cars is to air-condition them, with air-conditioning equipment using CFC gases as refrigerants. When the CFC refrigerants leak, the covalent fluorocarbon compounds (which are so stable at low levels in the atmosphere as to be relatively harmless to humans, unlike, say, ammonia) migrate slowly upwards through the stratosphere, where very energetic UV-C radiation from the sun eventually breaks them down. The Montreal Convention on CFCs therefore produces particularly acute current problems for hot developing countries. Freezing the status quo consumption of CFCs would not be just.

Developing and making more effective use of natural passive cooling techniques is currently especially important, as the only realistic current

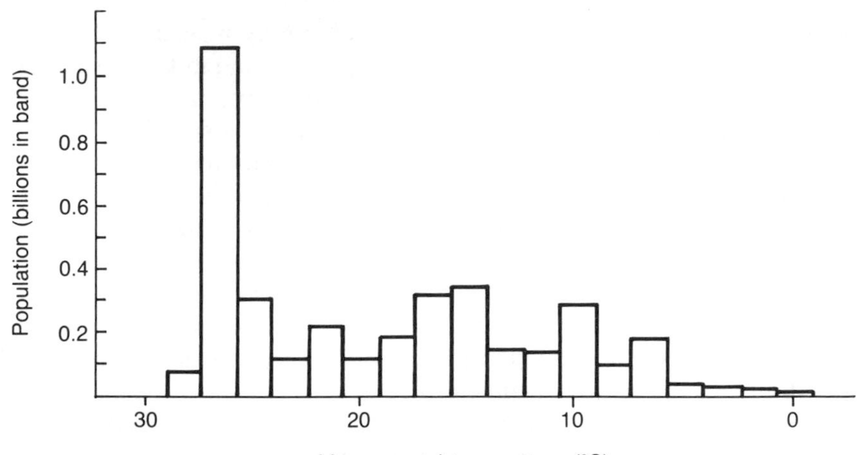

Figure 3.3 The concentration of population per 3°C temperature intervals versus mean annual temperature. (Source: McKay and Allsopp, 1980.)

alternative for promoting human comfort and health indoors under hot conditions involves using expensive, and seldom affordable, air-conditioning plant with CFCs as the current refrigerant gases. Human health can be very adversely affected by excessively hot conditions.

The old and young are especially vulnerable (WHO, 1989). In contemporary air-conditioned buildings, human cooling is achieved at the expense of adding to both global warming risks and to ozone destruction. The design problems of natural cooling have been addressed in detail in two projects, initiated by UNCHS Habitat (UNCHS, 1984; Page, 1989). These two projects have been specifically directed towards the needs of developing countries, and provide an international scientific basis for effective national action on passive solar heating and natural cooling. The evolution of simple design orientated methodologies for making reliable design attainable by ordinary designers, working in different climates, is critical for progress.

3.7 The scientific future

It seems likely that many clues to future progress on human solar energy applications may be found in nature. We are beginning, for example, to work on 'smart windows' with controllable wavelength dependent radiative properties, but 'smart leaves' have been around for a long time, accepting the photochemically useful energy, rejecting by reflection and transmission much of the near infra-red energy (useless for photosynthesis, which, if absorbed, would create, through heating, a greater water demand to make the plant more vulnerable to water shortages), incorporating protection against damaging UV radiation, providing a light controlled steerable surface, simultaneously capable of collecting and redistributing rain water to the roots, and logically constructed from materials in the immediate environment.

The big new scientific advances in solar energy applications during my working life seem to have come in the field of photovoltaics. This is primarily a success story in materials science backed by physical science. There are also the emerging advances in selective absorbing and transmitting surfaces, and new devices like electrochromic windows, again dependent on materials science. The advances in knowledge in the biological field have also been very impressive, but plants do not seem to have become any cheaper as energy sources, however, they are becoming more appreciated as a chemical materials resource. Their workings are certainly much better understood scientifically. However, a 0.75% level of energy efficiency is too low to make the impacts now needed on world energy supplies, and we must look to systems with much higher efficiencies to save both land and resources. This statement is not meant to undervalue in any way the importance of running biomass energy systems in parallel with basic food production cycles.

The recent advances in the manufacture of solar thermal power systems based on linear parabolic focusing reflectors is producing impressive falls in the costs associated with this route in areas favourable for focused solar energy (Tabor, 1989). The output per square metre of the most recent solar thermal plants exceeds by a factor of five the output of early plant. Current designs of linear parabolic collectors are producing peak energy at costs not far from the cost of conventional fuels, costed without their environmental charges. Looking at shortages of cropping land, we must look towards the deserts as unexploited and currently relatively unusable land areas. The Brundtland Report (1987) indicates 29% of the earth's land surface suffers slight, moderate or severe desertification; an additional 6% is classified as extremely desertified. If we could eventually extract and transport solar energy from the deserts, at 10% efficiency of collection, the production would vastly exceed the current world biomass production. Some of the problems lie with energy transport to the points of demand. The two promising routes are hydrogen and electricity.

I now have a strong suspicion it may be the turn of the chemists and biochemists to make the next rapid strides through new developments in the ways useful chemical reactions can be achieved in much more energy efficient ways by structuring the synthesis in logically constrained ways using 'chemical jigs'. One might broadly call the field chemical informatics. It embraces concepts like catalysis and enzymetic reactions, but probably a lot more. The chemists will have to match photochemical informatics with my main personal field of scientific interest, namely radiation environment informatics. It is clearly a wavelength dependent task. The energy can be supplied in a very clean environment. Only at this stage will we start to catch up with the 'smart leaf'. We must not value energy for its own sake, but for what we can do with it. A piece of oak in the hands of a craftsman may make a fine table or in the hands of a fool, a lump of firewood. Therefore we need to think more seriously about the sun in relation to problems of chemical synthesis, for example the production of nitrogenous fertilizers, or the provision of basic feedstocks for the chemical industries of the world.

3.8 Conclusion

The Brundtland Report (1987) has stressed the need for sustainable development. Mid–21st-century humans will wonder how, in a world where practically every natural activity is driven by energy, either directly or indirectly derived from the sun, 20th-century humans could so systematically overlook the facts of chemical and biological order implicit in the natural sun governed world, and become so trapped, in relation to the future of their species, by their waste cycles.

The challenges now faced in the field of human solar energy applications are clearly getting more complex, but the offsetting economic, social, environmental and health benefits are simultaneously becoming much more evident. It is therefore a particularly exciting time at which to be associated with solar energy. As one becomes older, the realization comes that one will not achieve one's dreams. If one is wise, and one's dreams are sound, one knows their achievement rests safely in the hands of the young.

References

Blaxter, K. (1986) *People, Food and Resources*, Cambridge University Press, Cambridge, UK.

Brundtland Report (1987) *Our Common Future*. Report of the World Commission on Environment and Development, New York.

Commission of the European Communities (1988) *Euroforum New Energies*. Proceedings of the International Congress, Oct. 1988, Saarbrucken, Germany, H. S. Stephens and Associates, Bedford, UK, Vol. 1–3.

Commission of the European Communities (1990) *European Reference Book on Daylighting*, CEC, Brussels.

Daniels, F. (1964) *Direct Use of the Sun's Energy*, Yale University Press, New Haven, USA.

Getman, F. H. and Daniels, F. (1943) *Outlines of Physical Chemistry – 7th edition*, Wiley, New York.

Hall, D. O. (1989a) The contribution of biomass to global energy use. *Biomass Journal*.

Hall, D. O. (1989b) Carbon flows in the biosphere: present and future. *Journal of the Geological Society, London*, **146**, 175–81.

Heede, R. E., Morgan, R. and Ridley, S. (1986) US Renewable energy policy: the hidden issues of energy. *Sunworld*, **10**, 36–42.

Hohmeyer, O. H. (1988) Macroeconomic view of renewable energy sources, in *Euroforum New Energies*, Proceedings of the International Congress, Oct. 1988, Saarbrucken, Germany. (eds Commission of the European Communities), H. S. Stephens and Associates, Bedford, UK. Vol. 1, pp. 12–18. Full report in Hohmeyer, O. H. (1988) *The Social Costs of Energy Consumption – External Effects of Electricity Generation in the Federal Republic of Germany*, Springer Verlag, Berlin.

McKay, G. A. and Allsop, T. (1980) The role of climate in affecting energy demand/supply in *Interactions of Energy and Climate* (eds W. Bach, J. Pankrath and J. Williams), D. Reidel Publishing Company, Dordrecht.

Page, J. K. (1989) *How to Prepare Locally Based Design Manuals for Passive Solar Heated and Naturally Cooled Buildings*. UNCHS Habitat, Nairobi.

Tabor, H. (1989) Solar Power – Report of World Energy Conference Committee on Solar Power, submitted to WEC meeting Sept. 1989, Montreal. World Energy Conference, London.

UNCHS (Habitat) (1984) Energy Conservation in the Construction and Maintenance of Buildings, Vol. 1. Use of Solar Energy and Natural Cooling in the Design of Buildings in Developing Countries. Report of the Expert Group meeting, Nairobi, 1983, Chairman J. K. Page, UNCHS, Nairobi.

UNEP (1987a) The Greenhouse Gases. UNEP/GEMS Environment Library No. 1, UNEP, Nairobi.

UNEP (1987b) The Ozone Layer. UNEP/GEMS Environment Library No. 2, UNEP, Nairobi.

UN Statistical Office (1988) Report on Ad Hoc Expert Group Meeting on Methods for the Collection and Compilation of Statistics on New and Renewable Sources of Energy, Rome, Sept. 29-Oct. 3 1986, UN Statistical Office, New York.

World Health Organization (1989) *Indoor Environment: Health Aspects of Air Quality, Thermal Environment, Light and Noise* (ed. J. K. Page), WHO, Geneva.

The greenhouse effect and future urban development

4

Ian Lowe

4.1 The future

When I was Acting Director of the Commission for the Future, I received many requests to make specific predictions about the future. The lesson of history is that people who make specific predictions about the future often finish up with egg on their face, even if it's in an area where they could be presumed to have some expertise. It has been said that it's difficult to make forecasts, especially about the future. There are two reasons for this: one related to information and the other more fundamental. First, we usually don't have enough information to make definitive predictions. At the peak of scientific hubris, there was a view that if only we knew the position and velocity of every particle we could chart the entire future of the universe through the laws of Newtonian mechanics. Of course, we don't have anything like that much information. However, even if we did have all that information and the capacity to process it, the more fundamental problem is that the future of the universe isn't deterministic in that simple mechanistic sense. Just as the world in which we live today has been significantly shaped by previous generations, so the future is being significantly shaped by individuals, groups and corporate entities, whether acting consciously or unconsciously.

Thus there's a fundamental sense in which we can't predict the future;

Global Warming and the Built Environment. Edited by Robert Samuels and Deo K. Prasad. Published in 1994 by E & FN Spon, London. ISBN 0 419 19210 7.

the future doesn't just happen according to some laws of physics, still less does it happen according to the 'laws' of economics, the currently popular form of metaphysics. The future is essentially a social construction. Indeed, many now would argue that knowledge in general is a social construct; at any given time there are a range of theories or a range of models which suit the world that we observe, and the choice between them is always partly a matter of social values.

The current dominance of technocracy in our society has been based on the implicit assumption that there is a technical view of the universe which is intrinsically correct, and before which all mere social constructs must stand aside. Our technical knowledge of the universe is also a social construct and subject to many of the same qualifications as our view of style or our view of aesthetics. Within a human lifetime, climatology has moved from being an observational science to a science which we understand in a primitive way with models. Now we are actually changing the climate, because the scale of the human population and the dimensions of our activities mean that we are now an active geological agent. In that sense, study of climate is now as much a branch of human activity as poetry or the performing arts.

4.2 Sustainable development

It is now clear to most reasonable people that we should think about our responsibility for the future. Thinking about the future leads to the conclusion that we should behave as if we intend to be permanent inhabitants of the planet, rather than as if we were visiting on a mission to loot, rape and pillage. That means we should try to move towards a pattern of development which is, at least in principle, sustainable. The phrase 'sustainable development' is currently on every lip in capital cities of the world, and even those who clearly don't understand it feel obliged to say that they're in favour of sustainable development, whatever it might be. Some even manage the mental gymnastics required to profess simultaneous allegiance to sustainable development and the free play of market forces!

Pearce has found more than 70 definitions of sustainable development. I think the best definition is the simple one of the Brundtland Report, the report of the World Commission on Environment and Development (WCED, 1987). It said that sustainable development is a pattern of activity that meets the needs of the present without compromising the ability of future generations to meet their needs. As a counter-point to the current tide of what is being called economic rationalism, which seems to be one of the great oxymorons of our time, it is salutary to note the comment in the Brundtland Report that unless economic decisions are **ecologically** rational we'll be unable to maintain, let alone improve, living standards.

In other words, while there may be some benefit from economic rationalism, it is more important to behave in a way that is ecologically rational, so that we can face the future with a measure of confidence.

For an activity to be sustainable it must not deplete natural resources significantly, it must not have serious impacts on the natural environment, and it mustn't prejudice social stability. The first two criteria are obvious; if we deplete resources or seriously alter the natural environment, we clearly prejudice the ability of future generations to meet their needs. The third criterion is important because it is easy to imagine policies that would slow down the rate of resource use or environmental impact, but which would be a clear threat to social stability. For example, it might be argued by those who have faith in market forces that the way to reduce transport fuel use in cities towards a sustainable level would be to increase the price to such prohibitive levels that many people would be unable to afford to drive long distances. As a general rule, those who live furthest from the centres of our cities travel the longest distances, have the least disposable income, are least likely to have public transport options and have the least efficient cars. So restricting transport by the price mechanism would have very serious consequences on social stability, particularly in the outer suburbs of our sprawling cities.

Similarly, it is possible to imagine a policy which would be ecologically sustainable within one region or even one nation, but which could bring that region or nation fundamentally into conflict with its neighbours. For example, if in Australia there was a long-term policy for coping with the run-down of oil production from Bass Strait it might be based on using the oil that is expected to be found under the Timor Gap.

Similarly, if Indonesia had a long-term strategy for meeting its liquid fuel needs it could well be based on using the oil that they expect to find under the Timor Gap. Indeed, the people of Timor seeking their independence from the colonial construct of Indonesia might also see the same oil resources as their potential economic salvation. It does not seem unnecessarily alarmist to suggest that conflict over oil could lead to friction or even war; there is a certain amount of empirical evidence suggesting that as a real possibility.

Therefore in aiming for sustainable development we need to be aware that this is a global issue, and we should take account of the global dimension. As the scale of the human population and the level of use of resources has expanded, environmental issues have been transformed from being local problems to regional issues, then to problems of global nature.

4.3 The greenhouse effect: IPCC recommendations

A recent authoritative statement of current scientific knowledge about the greenhouse effect was given by Working Group 1 of the Inter-Governmental Panel on Climate Change (IPCC, 1990). In adopting the recommendations of its scientific working group, the Panel was very careful to specify what is certain, what is calculated with confidence, what is predicted and what is uncertain.

It is certain that there is a natural greenhouse effect which keeps the earth warmer than it would otherwise be, that human activities are increasing the amounts of greenhouse gases in the atmosphere and that 'these increases will enhance the greenhouse effect, resulting on average in an additional warming of the earth's surface'.

It is calculated with confidence that carbon dioxide emissions are responsible for over half the enhanced greenhouse effect, and that overall emissions would need to be reduced to about 40% of the current level to stabilize the composition of the atmosphere. Based on current models, the Panel predicted that the rate of temperature rise will be greater than any experienced in the last 10 000 years; on current trends, about 0.3°C per decade.

If no action is taken to scale down the emissions of greenhouse gases, the mean estimate of increase is 1°C by 2025 and 3°C by the year 2100. It is estimated that there will be an average increase in sea level of about 6 cm per decade, mainly due to thermal expansion of water in the oceans, with some melting of land ice.

The Panel warned that these estimates could easily be out by as much as a factor of two in either direction, and said that various complicating factors mean that rates of increase will not be uniform. Some have argued that the uncertainties in these calculations mean that we need not worry. We should be clear what the uncertainty means. Things may turn out better than predicted, but it is as likely that things will be worse. In fact, the IPCC said explicitly that the unknown features are, in the collective judgement of that group, likely to make things worse than we fear, rather than making them better than we hope. The sombre warning was that 'the complexity of the system means that we cannot rule out surprises'.

Among other predictions, it is believed that there will be an increased frequency of extreme events, such as heavy rainstorms and flooding of low-lying areas. Temperature increase is also thought likely to bring tropical cyclones further south than in recent years, increasing the risk of storm surges affecting coastal areas. The judgement of the IPCC is that the observations of climate change provide strong support for the view that global warming is already detectable. The increase in average temperatures this century of about 0.5°C, the increase in sea level of 10–20 cm and the recent experience of the five warmest years ever recorded being in the 1980s, are all observations which strengthen the conviction that we are seeing real change. The IPCC note, however, that it will probably require another decade of measurement to be able to say with complete assurance that enhancement of the greenhouse effect is being measured.

Despite that note of caution, the broad conclusion of the IPCC was that 'in the absence of major preventative and adaptive actions by humanity, significant and potentially disruptive changes in the earth's environment will occur . . . The world community recognises the need to undertake certain actions to reduce and mitigate the impact of climate change.'

4.4 Sources of carbon dioxide

Carbon dioxide is produced by the burning of compounds which contain carbon: coal, oil, gas and wood are the principal examples. Carbon dioxide is absorbed from the air by growing plants, which use the energy of sunlight to combine carbon dioxide with water from the soil to produce carbohydrates. For this reason, the loss of vegetation contributes doubly to the greenhouse effect. Clearing plant material causes the stored carbon to be returned to the air as carbon dioxide. It also reduces the amount of plant material available to extract carbon dioxide from the air for new growth.

Worldwide, about 6000 million tonnes of carbon are added to the atmosphere each year by the burning of fossil fuels. The estimates of the additional contribution from the clearing of vegetation are given by Smith, Thambimuthu and Clarke (1991) as ranging from 800 to 2400 tonnes of carbon per year; most estimates are in this range, which suggests that the clearing of vegetation accounts for about 20% of the increased carbon dioxide. Thus the clearing of forest and bush-land makes a significant contribution to the problem of global climate change.

4.5 Carbon dioxide targets

The Australian government has adopted as an interim target the goal suggested by a 1989 conference in Toronto, to reduce the level of emissions of greenhouse gases to 20% below the 1988 level by the year 2005. The scientific assessment suggests this to be a minimum target, given that the reductions needed to stabilize the composition of the atmosphere are actually much greater. The IPCC said that it would require a 60% reduction in emissions of carbon dioxide and other long-lived greenhouse gases to stabilize their levels in the atmosphere. It is therefore urgent to devise practical ways of reducing our impact on the atmosphere.

It is important to be aware of the scale of the problem. A further complexity is that Australians currently emit more carbon dioxide per head than most other countries. Given that there is a reasonable expectation in Third World countries that they will improve their standard of living, if not to that now enjoyed by the industrialized countries at least to something closer to human dignity that the conditions under which many people now live, the level of fuel carbon emissions which will eventually be seen as reasonable for Australia is probably about 20% of the present value. Thus dealing with global warming will not simply require minor cosmetic changes, but a fundamentally different approach to the use of energy. That is an inescapable conclusion.

There are three broad approaches to the task of reducing the emissions of carbon dioxide. The level of energy can be reduced by improved efficiency or conservation measures, carbon-based fuels which emit less carbon dioxide per unit of energy can be used, and increased use of non-carbon fuels can be made. These are not mutually exclusive; indeed, a sensible strategy to reduce emissions will include elements of all three approaches.

In 1990 the Australian government set up a range of working groups to explore aspects of ecologically sustainable development; the greenhouse effect was one of the key issues for these working groups. The reports of the Ecologically Sustainable Development Working Groups can be seen as a timid first step in the direction of developing a strategy to meet our needs without unacceptable long-term effects (ESD, 1991). It was concluded that it will be difficult to reduce the rate of burning fuels by 20% by the year 2005 and argued, more dubiously on the basis of naïve economic notions, that to do so might cause considerable economic hardship. This questionable conclusion has been used in a determined campaign to try to persuade the government to back down on its commitment to the Toronto target, although impartial studies suggest that the economic benefits from use of more efficient technologies would be considerable.

4.6 Effects on the built environment

Since anybody who designs, commissions or constructs a building probably expects it still to be there in 30 or 40 years time, one of the factors which ought now to be considered is the climate changes which are possible in its lifetime. Thus design should take account of possible rises in temperature, increases in rainfall, changes in sea level and increases in the frequency of extreme events.

It should also be considered how urban form contributes to the problem. There are several ways in which the construction of buildings affects the atmosphere. One is construction energy: the building materials used require fuel and resources for their production. I calculated the fuel energy required for a typical domestic house in Brisbane; there is more than a factor of two difference between the fuel energy required for a traditional 'Queenslander' made of timber, with verandahs and a galvanized iron roof, and the sort of brick box which is often and inappropriately being built in the Brisbane suburbs, with a concrete slab, brick walls and a tiled roof. It takes more than twice as much fuel energy for the brick house, which is usually less suitable for the Brisbane climate than the more traditional vernacular housing.

Secondly, fuel is required for the operation of buildings. It is used to produce an acceptable climate internally, for the provision of lighting, to operate lifts and other machinery and so on. Resources are also used to supply water to the buildings.

There are some obvious principles that ought to be followed in designing and constructing buildings, for example, awareness of the effect of design and orientation on the needs for climate modification after construction. The principles of designing and orienting buildings to take advantage of the sun in winter, and not be inconvenienced by it in summer, have been known for about 2500 years. While it would be hasty to rush new ideas into practice, it is inexcusable for professional people to be unaware of the influence of orientation and design on the solar gain of buildings and the consequent need for climate modification. Many buildings in cities combine so many features of bad design that they might have been sponsored by the electricity industry to boost flagging sales.

Secondly, we need to be aware of the operating efficiency of equipment. Recent analysis shows that there is no systematic price penalty for efficient appliances; the correlation between purchase price and operating efficiency is essentially zero. For appliances of a given capacity, there is typically about a factor of two difference between the most efficient and the least efficient. There is no economic reason to use less efficient equipment. Indeed, in the case of some appliances the less efficient equipment is actually significantly more expensive.

In designing buildings, there should be an awareness of the ways in which control systems and organization can economize on fuel use. For example, it is not unusual in large office buildings for an entire floor to be lit if one person is still working. We also need to be aware of the ways in which human behaviour affects energy use. It is bizarre that Australian men who work in offices wear jackets and ties in summer. This form of dress is arguably defensible in the European or North American winter, but is absurd in the sub-tropical summer. It often leads to the use of megawatts of cooling energy to refrigerate whole buildings so men can work dressed in this unsuitable style. Women, who usually dress more sensibly in summer, often have to take warm clothing to work in Brisbane in January so that they don't suffer from the cold. It should be recognized that in such a sub-tropical climate business people should dress as their counterparts do in Kuala Lumpur or Manila, rather than as those in London or Boston.

Finally, in constructing buildings we need to think about water needs, and in particular about the opportunity for using 'grey water'. Water which has been used, for example, for washing hands is quite suitable for purposes which don't require sterile and purified water, such as flushing toilets. It is absurdly wasteful to spend a fortune sterilizing and purifying water which will be used to flush toilets or water gardens. There are opportunities to make use of water which is only mildly dirty for those sorts of activities. This principle is now being applied to some new developments in Australia; it should be a general rule.

A final consideration in the design and use of buildings is lighting. We should consider the appropriate levels of lighting for particular purposes, as well as taking advantage where possible of opportunities to use natural light. Where it is essential to supplement daylight the technology for supplying light should be carefully considered. The fluorescent tube is better, and the compact fluorescent tube is better still. Using more efficient lights also reduces the need for air-conditioning, as about half of the load in a large building stems from the waste heat of lights. Since the more efficient lights also last longer, they typically also produce maintenance savings in commercial buildings.

4.7 Urban form

Attention should be paid to the way the lay out of buildings constrains the pattern of movement in cities. The current pattern of urban transport is characterized by high energy, high pollution levels, a high level of death and injury, and a high level of social dislocation. Road 'accidents' are the largest single cause of lost life years among adult males in Australia, and the problem is so depressingly predictable that it is referred to as the road toll, as if it were the inevitable price of using cars. The current pattern of urban transport fails on all three criteria of sustainability. It is rapidly using fuel resources in a way which is not sustainable; it is producing pollution levels which are often unacceptable and it contributes to the break down of the social fabric. A sustainable transport system would involve much less use of cars, much more use of public transport, and much more cycling and walking.

While the world as a whole is not generally short of resources, there is one particular problem. For the best part of 30 years now, it has become steadily more difficult to find oil (Lowe, 1986). It is taking more and more exploration to find new reserves. We are now getting oil from such unlikely places as the North Sea and Alaska. There is pessimism about the chance of ever again finding the sort of massive deposits which have fuelled the expansion in transport since 1950. The current rate of using

Table 4.1 Mode of travel by commuters (%)

Region	Public transport	Pedestrians or cyclists	Private transport
USA	12	5	83
Australia	19	5	76
Europe	35	21	44
Asia	60	25	15

Source: Newman and Kenworthy (1989)

petroleum fuels does not just threaten the global atmosphere; it also risks depleting the oil fields within the life-time of today's young people. Production from the USA and the former USSR is in rapid decline, making the world more dependent on oil from the Middle East (Flavin, 1992).

In terms of carbon dioxide emissions, road transport accounts for about four times as much as rail, sea and air together, with the vast majority of the emissions from road transport occurring in urban areas (Watson and Watson, 1990). Other exhaust gases, such as carbon monoxide and oxides of nitrogen and sulphur, produce serious air pollution problems in cities (Simpson and Lowe, 1991).

The basic reason is that motor fuel use in Australian cities, as shown by Newman and Kenworthy (1989), is about double the average level for the cities of western Europe and about four times the level for prosperous Asian cities such as Singapore and Tokyo. Only the USA has urban areas in which fuel use is greater. Table 4.1 shows the scale of the problem. This table shows that Australian commuters are almost twice as likely as Europeans and five times as likely as those in affluent Asian cities to use a car. By contrast, European and Asian commuters are twice and three times as likely, respectively, to use public transport. Only 5% of Australian urban commuters walk or cycle, whereas the average figure in Asian cities is 25% and in western Europe 21%. One reason for this is that Australians make greater provision for cars. For example, road space, in metres per person, is even more generous than in the USA, and about four times as great as that provided in the cities of Europe. As a result, cars in Australian cities travel faster than those in European or Asian cities, yet it is constantly suggested that roads have to be widened and improved to allow a freer flow of cars. Houses, shops or even graveyards that stand in the way are bulldozed to allow cars to move more rapidly from one traffic jam to the next.

Those who put their trust in economic forces suggest that this behaviour pattern could be changed by something like a carbon tax. The sort of carbon tax that has been discussed is something like $US50 a tonne of carbon emitted. To put that in perspective, the current taxes on motor fuels in OECD countries range from about $120 a tonne to about

Table 4.2 Energy for modes of transport

Mode	Energy (J/passenger/km)
Car (one occupant)	4 800
Transit bus	2 400
Transit rail	2 300
Walking	260
Cycling	90

$1400 a tonne. The current effective carbon tax on motor fuels in Australia is about $250 a tonne, for example, so a carbon tax of $50 a tonne could hardly be considered a penal imposition. If applied globally, however, such a tax would raise something like $US300 billion a year. This is a very large sum of money, which could massively redirect transport use if it were used to provide alternatives to current practices.

It has also been suggested that the users of fossil fuels should pay full costs, including the cost of depletion of resources, the environmental cost of their use, the military cost for protecting access to resources and so on. In protecting access to oil from Kuwait in 1990–91, the US military spent much more in one year than has been spent throughout all human history on trying to develop alternative sources of energy. While it is sometimes said that we cannot afford to develop alternative sources of energy based on renewable forms, the reason, clearly, is that we have chosen not to make that a priority.

One part of the solution would be to make more use of bicycles for urban transport. There are four sound reasons for doing this: they save fuel, they reduce emissions, they improve community health and they bring social benefits. The amazing energy-efficiency of modern bicycles is clearly a strong point. Table 4.2 shows the energy required for various modes of transport (data adapted from M. Lowe, 1989). Cycling uses one-third the energy of walking, about one twenty-fifth the energy of public transport and about one fiftieth the energy of an average private car. It should also be noted that the cyclist runs on renewable fuel rather than on petroleum.

The point about emissions is obvious. An average Australian car releases about seven tonnes of carbon dioxide each year. When I sold my car and our family became a zero-car, four-bicycle operation, that reduced the Australian emission of fuel carbon. Each replacement of one commuter trip by a cycle trip makes a contribution, since each saving of a litre of motor fuel is equivalent to about 3 kg less carbon dioxide emitted (I. Lowe, 1989).

The health benefits occur at two levels. Only about 5% of adult Australians exercise sufficiently to maintain basic heart–lung fitness, while a much greater percentage are over-weight to an unhealthy extent (Hetzel and McMichael, 1987). Increasing the number of adults cycling would increase the fitness of the population. Despite the poor facilities for cycling in Australia, it is still safer than other modes of private transport for the group most at risk of road accidents in Australia, young men. In terms of distance travelled, a cyclist is about half as likely to have a fatal accident as a car driver, about one-third as likely as a pedestrian and less than a tenth the risk of a motor cyclist. Since a significant fraction of the fatal accidents involving cyclists also involve other vehicles, better facilities for cycling would make it an even safer means of transport (Robinson, 1991).

4.8 The political imperative

The broad perspective we need is in the Brundtland Report (WCED, 1987), which argues that meeting the essential needs of the world community can be consistent with economic growth, provided that the type of growth reflects the broad principle of sustainability and non-exploitation of others. It requires that we meet human needs by increasing productive potential and ensuring equitable opportunities for all. As was recently said:

> The most important question we now face is whether we collectively have the foresight and courage to make the necessary changes. Meeting the challenges will require unprecedented cooperation and vision. There is no quick technological fix. The critical issue is and will always remain, the need to plan for a more secure and sustainable world. It will require a long-term commitment from all of us as individuals and a fundamental change in our attitude to the use of resources.

That statement was made by an arch-pragmatist, Senator Graham Richardson (1990), in his capacity as Australian Minister for the Environment. The fact that we are now getting at least a greener shade of rhetoric, even if not a greener shade of action, shows that politicians are becoming increasingly aware that the ordinary members of the community want these long-term issues to be taken into account. It will be increasingly expected of professional people in the future that they don't only understand the laws of mechanics and the fuzzier laws of economics, but also the laws of ecology. These were phrased by Commoner (1971) in these terms:

- everything is connected to everything else;
- everything must go somewhere;
- nature knows best;
- there is no such thing as a free lunch.

We need to be aware of the ecological implications of our actions. The future doesn't just happen. We are all engaged consciously or unconsciously, actively or inactively, jointly and severally, in the exercise of creating the sort of future we will all have. We are much more likely to have the sort of future we want if we aim purposively for it, rather than follow technical or economic creeds in the hope that things will just turn out right.

This idea can be illustrated by an analogy. Very few people go on holiday by loading their family and clothes into a car and driving randomly around the road system, hoping they find somewhere interesting before they run out of petrol. Most people going on holiday have at least a general plan of where they want to go and what they want to do. They may retain the flexibility to respond to new problems or new opportunities by varying their plan, but they usually have at least a general scheme. The future of our urban areas deserves at least as much consideration as a family holiday. In the same terms, we should have at least a general picture of where we want to go and what sort of future we want to have. If we have a general picture of that type, we're much more likely to achieve the sort of future we want than if we just hope it will happen.

There is a growing public mood for change. A regular poll conducted by the *New York Times* and *CBS News* asks Americans if they agree with the statement that 'Protecting the environment is so important that requirements and standards cannot be too high, and continuing environmental improvements must be made regardless of cost.' This quite strong statement was being asked in the early 1980s, when there was a small majority in favour. The level of support has grown steadily throughout the 1980s. By mid–1989 over 80% of Americans sampled agreed with this statement, with only 13% disagreeing (I. Lowe, 1989).

Dr Mostafa Tolba (1989) of the UN Environment Program has said that people the world over want their political, scientific and industrial leaders to be doing far more about ecological issues. He said that people want action, and show every sign of being prepared to pay the price of acting to secure a better future, but governments are lagging behind public opinion. There is an obvious reason for this delay. Finding a mode of human development that will be sustainable is a very complex task. It requires an understanding of many interacting aspects of the problem. It is a great challenge to all of us, but one which our responsibility to all future generations requires us to address.

References

Commoner, B. (1971) *The Closing Circle*, Knopf, New York.

ESD Working Groups (1991) Final Report – Executive Summaries, AGPS, Canberra.

Flavin, C. (1992) Building a bridge to sustainable energy, in *State of the World 1992* (eds L. R. Brown *et al.*), Worldwatch Institute, Washington DC, USA.
Hetzel, B. and McMichael, A. (1987) *The LS Factor*, Penguin, Ringwood.
IPCC (1990) *Climate Change, The IPCC Scientific Assessment*, Cambridge University Press, Cambridge, UK.
Lowe, I. (1986) Australian Energy Demand 1980–2030, NERDDC End of Grant Report, Department of Resources and Energy, Canberra.
Lowe, I. (1989) *Living in the Greenhouse*, Scribe Publications, Newham.
Lowe, M. (1989) The Bicycle: Vehicle for a Small Planet, Worldwatch Paper No. 90, Worldwatch Institute, Washington DC, USA.
Newman, P. W. G. and Kenworthy, J. (1989), *Cities and Automobile Dependence*, Gower Publications, Aldershot, UK.
Pearce, D., Markandya, A. and Barbier, E.B. (1989) *Blueprint for a Green Economy*, Earthscan Publications, London, pp. 1–27.
Richardson, G. (1990) Greenhouse – the challenge of change, in *Greenhouse, What's to be Done?* (ed. K. Coghill), Pluto Press, Sydney, pp. 67–76.
Robinson, B. (1991) *Bicycle Usage and Safety in Australia*, Bicycle Federation of Australia, Canberra.
Simpson, R. and Lowe, I. (1991) Environmental Impact Review, South-East Queensland Passenger Study, Queensland Department of Transport, Brisbane.
Smith, I. M., Thambimuthu, K. V. and Clarke, L. B. (1991) *Greenhouse Gases, Abatement and Control: the Role of Coal*, IEA Coal Research, London.
Tolba, M. (1989) Speech on World Environment Day, United Nations Environment Program, Nairobi.
Watson, H. C. and Watson, C. R. (1990) Near and long term prospects for the reduction in the road transport contribution to greenhouse gases, in *Greenhouse and Energy* (ed. D. J. Swaine), CSIRO Publications, Melbourne, pp. 320–30.
WCED (1987) *Our Common Future*, Oxford University Press, Oxford, UK.

Urban design, transportation and greenhouse*

5

Peter Newman

5.1 Introduction

Urban design, or the shape of our cities, has not featured strongly in the greenhouse debate. There is rather, a natural tendency for scientists and engineers to favour technological assessments and solutions. It is rarely considered that the placement of buildings and how compactly they are grouped will make a major impact on transportation and thus the production of greenhouse gases.

There are few statistical studies of the way that urban design varies around the world and hence a study was initiated in the 1980s to collect and analyse data on 31 cities globally. This study, funded by the Australian Government, took the best part of a decade to complete. The results of this survey are presented here to suggest ways that urban design can modify transportation and contribute to reductions in greenhouse gases.

5.2 Global cities and a greenhouse comparison

In Table 5.1 31 cities are outlined in terms of their per capita CO_2 emissions from the transport fuels gasoline, diesel, and electricity (some LPG

* This chapter is an expanded version of a paper published in *Futures*, May, 1991.

Global Warming and the Built Environment. Edited by Robert Samuels and Deo K. Prasad. Published in 1994 by E & FN Spon, London. ISBN 0 419 19210 7.

Table 5.1 Carbon dioxide per capita from transport in 31 world cities, 1980

City	Motor vehicles Gasoline (tonnes)	Motor vehicles Diesel (tonnes)	Total (%)	Public transport Diesel (tonnes)	Public transport Electricity (tonnes)	Total (%)	Total (tonnes)
Houston	5.14	0.65	100	0.02	–	<1	5.81
Phoenix	4.82	0.61	100	0.01	–	<1	5.45
Detroit	4.55	0.21	99	0.03	–	1	4.79
Denver	4.38	0.95	99	0.04	–	1	5.36
Los Angeles	4.03	0.44	99	0.04	–	1	4.52
San Francisco	3.82	0.67	97	0.06	0.07	3	4.62
Boston	3.74	0.69	99	0.03	0.03	1	4.4
Washington	3.54	0.15	97	0.05	0.06	3	3.79
Chicago	3.33	0.26	96	0.08	0.05	4	3.72
New York	3.04	0.43	89	0.04	0.38	11	3.88
US average	**4.04**	**0.51**	**98**	**0.04**	**0.06**	**2**	**4.64**
Perth	2.25	0.48	98	0.06	–	2	2.78
Brisbane	2.11	0.40	97	0.06	0.02	3	2.59
Melbourne	2.01	0.31	95	0.02	0.10	5	2.44
Adelaide	1.99	0.37	97	0.07	<0.01	3	2.43
Sydney	1.93	0.39	94	0.05	0.09	6	2.46
Australian average	**2.06**	**0.39**	**96**	**0.05**	**0.04**	**4**	**2.54**
Toronto	2.40	0.84	94	0.07	0.14	6	3.45
Hamburg	1.15	0.41	92	0.05	0.09	8	1.69
Frankfurt	1.11	0.39	89	0.01	0.18	11	1.69
Zurich	1.08	0.28	92	0.01	0.23	8	1.47
Stockholm	1.07	0.38	84	0.06	0.19	16	1.72
Brussels	1.02	0.46	85	0.04	0.22	15	1.74
Paris	0.97	0.34	89	0.02	0.14	11	1.47
London	0.86	0.36	88	0.04	0.14	12	1.39
Munich	0.85	0.30	86	0.02	0.16	14	1.33
West Berlin	0.78	0.28	87	0.05	0.11	13	1.22
Copenhagen	0.79	0.32	85	0.06	0.13	15	1.30
Vienna	0.70	0.08	82	0.02	0.15	18	0.95
Amsterdam	0.50	0.21	78	0.04	0.16	22	0.91
European average	**0.91**	**0.32**	**88**	**0.04**	**0.16**	**12**	**1.40**
Tokyo	0.76	0.32	83	0.02	0.20	17	1.30
Singapore	0.41	0.24	86	0.11	–	14	0.76
Hong Kong	0.14	0.18	82	0.05	0.02	18	0.39
Asian average	**0.44**	**0.25**	**84**	**0.06**	**0.07**	**16**	**0.82**

Derived from data in Newman and Kenworthy (1989); rounding errors mean that components of Table 5.1 do not always add to the totals. Conversion of energy use to CO_2 emissions is as follows; 69 g CO_2 per MJ gasoline and diesel; 60 g CO_2 per MJ LPG; 282 g CO_2 per MJ electricity. The electricity conversion assumes coal fired power stations (94 g CO_2 per MJ) and 2/3 of coal wasted due to power plant inefficiency and distribution losses; LPG is added to gasoline as it is mostly used by passenger motor vehicles; assumptions needed to be made about diesel use in Amsterdam (passenger car fuel data were increased to account for commercial traffic) and in Hamburg, Frankfurt and Munich (the same proportion of gasoline to diesel as in the other European cities was assumed).

is also included). The results are ranked in order of gasoline consumption since this is the major source of transport greenhouse gases. (All the other tables in this chapter containing data from world cities are also in that format.) The data show that US cities produce on average 4.64 tonnes of CO_2 per capita from transport compared to 2.54 tonnes in Australian cities, 1.40 tonnes in European cities and 0.82 tonnes in Asian cities. This is an enormous variation, much greater than could be explained by obvious economic factors. The following section sets out the apparent causes for these variations.

The breakdown by fuel type shows that gasoline is by far the biggest contributor to transport greenhouse gases, and this is most marked in US and Australian cities where the automobile is extensively used. Where cities are more public transport oriented, diesel and electricity use becomes much more significant – the percentages of different fuel types used in the Tokyo transport system, for instance, are 53% gasoline, 33% diesel and 15% electricity; whereas Houston percentages are 89% gasoline and 11% diesel.

The breakdown between private and public transport shows an overwhelming proportion of greenhouse gases derive from private usage. Public transport contributes only 2% of transport greenhouse gases in US cities, 4% in Australian cities, 12% in European cities, and 16% in Asian cities.

Diesel use has a remarkably uniform pattern across the cities. Hamburg and New York, for example, have almost the same per capita diesel use, indicating the similar dependence that most cities now have on the light van and truck for urban freight movement. However, the real difference between the cities is the comparative use of gasoline and electricity.

Table 5.1 assumes that the source of electricity consumed by electric transport is coal-fired power plants. Despite coal-based electricity being some four times worse than gasoline in terms of CO_2 produced per MJ of useful transportation energy it does not mean cities with electric transportation are worse greenhouse gas offenders; in fact the reverse is the case. This is because of the nature of the technology and the effect of either the car or the train/tram on the city. This urban design difference is fundamental to the thesis being presented here. It is an important factor in the greenhouse debate where coal is considered to be so much more damaging; even if coal is used to provide an electric train or tram system then the city will still produce lower levels of greenhouse gases overall. The mechanism for this is set out below.

In addition, the advantages of electric-based transit become far greater in terms of limiting greenhouse effects and acid rain, when the primary source of power is a renewable fuel. This will become a far more significant option in the decades ahead as oil prices continue to rise and a diversity of renewable fuels begins to feed into our electricity grid systems. In short the data show that cities which are based on electric public transportation systems have considerably reduced oil dependence and, consequently, reduced greenhouse gas emissions.

5.3 Transportation orientation and infrastructure

Table 5.2 sets out the proportion of transportation by non-automobile modes – public transit and cycling/walking – and a few transportation infrastructure parameters. The strong link between transit and cycling/walking is emphasized in these data. Rebuilding our cities to reduce oil dependence by improving electric transit appears to involve a parallel improvement in cycling and walking facilities.

The central conclusion in Table 5.2 is that those cities with the highest transit utilization (and these are virtually all rail-based systems) tend to have the lowest greenhouse gases emissions. This is an important finding because it is sometimes claimed that if we are going to use more transit then it is not much better than the automobile when it comes to emissions per capita. However, this is because simplistic assumptions are made about changing from one mode to the other. It is not just a matter of one trip by car being replaced by one trip on transit. Once a mode is provided that can adequately compete with the automobile then a range of changes occur, including:

- the elimination of some car trips because people are seen, goods are purchased and places are visited as part of a single transit trip, since transit stations invariably have a range of other urban facilities closely associated or nearby;
- more walking and cycling are engaged in as dependence on the automobile is reduced (see Table 5.2);
- land uses are modified within range of this new form of transportation (see Table 5.3).

Only by considering these urban design-related changes is it possible to appreciate the full extent of the potential benefits of rebuilding our cities around non-automobile modes. Holtzclaw (1990) argued, in an analysis of transportation and land-use patterns in San Francisco, that there is a 10:1 leverage when a transit trip replaces a car trip, i.e. 10 passenger km of car travel are replaced by 1 passenger km of transit. This microlevel principle is consistent with our data, although we would suggest that the leverage is more like 3.5:1. Thus, in those cities which have greater transit reliance total travel is considerably less than a modal shift alone would indicate. For example, Houston has 0.8% of its total travel of 16 100 passenger km per capita by transit. If Houston puts in an extensive rail service, the total fuel saved would be more than the savings from

Table 5.2 Non-automobile transportation patterns and provision for the automobile in 31 cities, 1980

City	Public transport (% of total pass. km of travel)	Bicycle and walking (% of workers for journey to work)	Road availability (m/ person)	Parking availability (per 1000 CBD workers)	Average speed of traffic (km/h)
Houston	0.8	2.8	10.6	370	51
Phoenix	0.5	3.2	10.4	1 033	42
Detroit	0.8	2.8	5.8	473	44
Denver	1.8	5.3	9.4	498	45
Los Angeles	2.7	4.2	4.5	524	45
San Francisco	6.6	5.5	4.9	145	46
Boston	4.0	9.8	5.2	322	39
Washington	5.0	5.2	5.1	264	38
Chicago	8.0	6.2	5.0	91	41
New York	14.1	8.1	4.7	75	35
US average	**4.4**	**5.3**	**6.6**	**380**	**43**
Perth	4.9	4.0	13.3	562	43
Brisbane	6.0	5.3	6.9	268	48
Melbourne	7.1	5.7	7.9	270	48
Adelaide	5.8	5.8	9.1	380	43
Sydney	13.8	5.4	6.2	156	39
Australian average	**7.5**	**5.2**	**8.7**	**327**	**44**
Toronto	16.7	5.8	2.7	198	–
Hamburg	17.0	15.3	2.2	149	30
Frankfurt	20.1	27.0	2.0	242	30
Zurich	22.9	21.0	2.6	140	36
Stockholm	24.4	20.0	2.3	153	30
Brussels	19.7	15.6	1.7	186	–
Paris	30.3	23.8	0.9	201	28
London	27.8	23.0	1.9	130	31
Munich	23.3	20.0	1.7	285	35
West Berlin	32.1	15.0	1.5	438	28
Copenhagen	21.0	32.2	4.3	212	45
Vienna	30.0	14.7	1.7	190	30
Amsterdam	28.9	28.0	2.1	208	39
European average	**24.6**	**21.3**	**2.1**	**211**	**30**
Tokyo	63.4	24.9	1.9	66	21
Singapore	52.1	15.8	1.0	97	30
Hong Kong	76.9	34.5	0.2	37	21
Asian average	**64.1**	**25.1**	**1.0**	**67**	**24**

transferring these few trips onto trains. These urban design factors must be considered when calculating potential savings.

Two relationships apparent in Table 5.2 have important implications:

- The cities with highest CO_2 emissions have the greatest provision for the automobile in terms of roads and car parking.

This would suggest that if we are seriously attempting to reduce automobile dependence, we must start by restricting the amount of resources we put into facilities for the automobile. This will free up an enormous source of capital to assist in rebuilding the city.

- The cities with lowest CO_2 emissions have the slowest moving traffic.

This finding contradicts the theory that free-flowing traffic saves fuel and reduces emissions due to greater vehicle efficiency. A more realistic explanation is that congestion does indeed reduce vehicle efficiency but it also makes other transport modes more attractive, and, most importantly, tends to limit city sprawl thus reducing travel distances. We have elaborated on these arguments in detailed studies undertaken in Perth (Newman and Kenworthy, 1988). The findings have been adopted in the USA where the Environment Protection Agency has now requested that highway design policies consider the effect of increased vehicle miles of travel (VMT) on emissions. This is an important step in recognizing the need for cleaner cities not just cleaner cars, and that urban design plays a central role in the management of urban transportation. These realizations are also now apparent in the new Clean Air Act in the USA which makes urban design a component of transportation policy for the first time.

In sum, electric rail-based cities are clearly the way we should be heading in future. Central to this concept is the role of rail in urban design.

5.4 Urban design

Table 5.3 shows the land-use patterns in the 31 cities. They indicate that the automobile-based cities with high CO_2 emissions have the lowest population densities. In contrast, the higher density, rail-based cities have low CO_2 emissions. This applies to overall density and the density of population in the central city, inner and outer areas. Job densities follow a similar pattern, except in US and Australian cities where high concentrations of employment opportunities are found in central (CBD) areas.

A large disparity between where people live and where they work means that trip distances are longer in US and Australian cities than in European cities. However, the high density of jobs located in central areas of the former is also the backbone of the public transportation

Table 5.3 Urban form in 31 cities in 1980 (persons and jobs per hectare)

City	Total density		Central city density		Inner city density		Outer area density	
	Pop.	Jobs	Pop.	Jobs	Pop.	Jobs	Pop.	Jobs
Houston	9	5	5	443	21	26	8	3
Phoenix	8	4	17	67	19	24	8	4
Detroit	14	6	11	306	48	20	11	5
Denver	12	8	18	263	19	17	10	5
Los Angeles	20	11	29	472	30	14	17	9
San Francisco	15	8	90	713	59	48	13	5
Washington	13	8	8	584	44	38	11	6
Boston	12	6	125	383	45	33	10	4
Chicago	18	8	16	938	54	26	11	5
New York	20	9	217	828	107	53	13	6
US average	**14**	**7**	**54**	**500**	**45**	**30**	**11**	**5**
Perth	11	5	8	121	16	15	10	3
Brisbane	10	4	15	346	19	16	9	3
Melbourne	16	6	25	647	29	40	16	4
Adelaide	13	5	8	251	19	25	12	4
Sydney	18	8	11	434	39	39	16	5
Australian average	**14**	**6**	**13**	**360**	**24**	**27**	**13**	**4**
Toronto	40	20	25	757	57	38	34	14
Hamburg	42	24	26	407	88	106	35	12
Frankfurt	54	43	65	389	63	74	49	25
Zurich	54	33	44	422	79	66	42	17
Stockholm	51	34	97	280	58	62	46	16
Brussels	67	42	74	592	101	85	50	19
Paris	48	22	235	400	106	60	26	8
London	56	30	66	397	78	62	48	19
Munich	57	34	111	231	159	192	48	21
West Berlin	64	27	133	333	84	46	57	20
Copenhagen	30	16	85	325	59	38	24	11
Vienna	72	38	65	403	133	113	59	23
Amsterdam	51	23	108	153	83	46	32	43
European average	**54**	**31**	**92**	**361**	**91**	**79**	**43**	**20**
Tokyo	105	66	82	477	153	114	58	20
Singapore	83	37	204	339	202	–	63	–
Hong Kong	293	110	160	1 259	1 037	478	224	66
Asian average	**160**	**71**	**149**	**692**	**464**	**296**	**115**	**44**

system in these cities. Thus, simply moving jobs from central areas to the suburbs is not going to guarantee any less travel overall – unless they are located proximate to high density areas and thus maintain transit viability.

The effect of density is fundamental to automobile use. Figure 5.1 shows that this is an exponential relationship. This can be seen just as

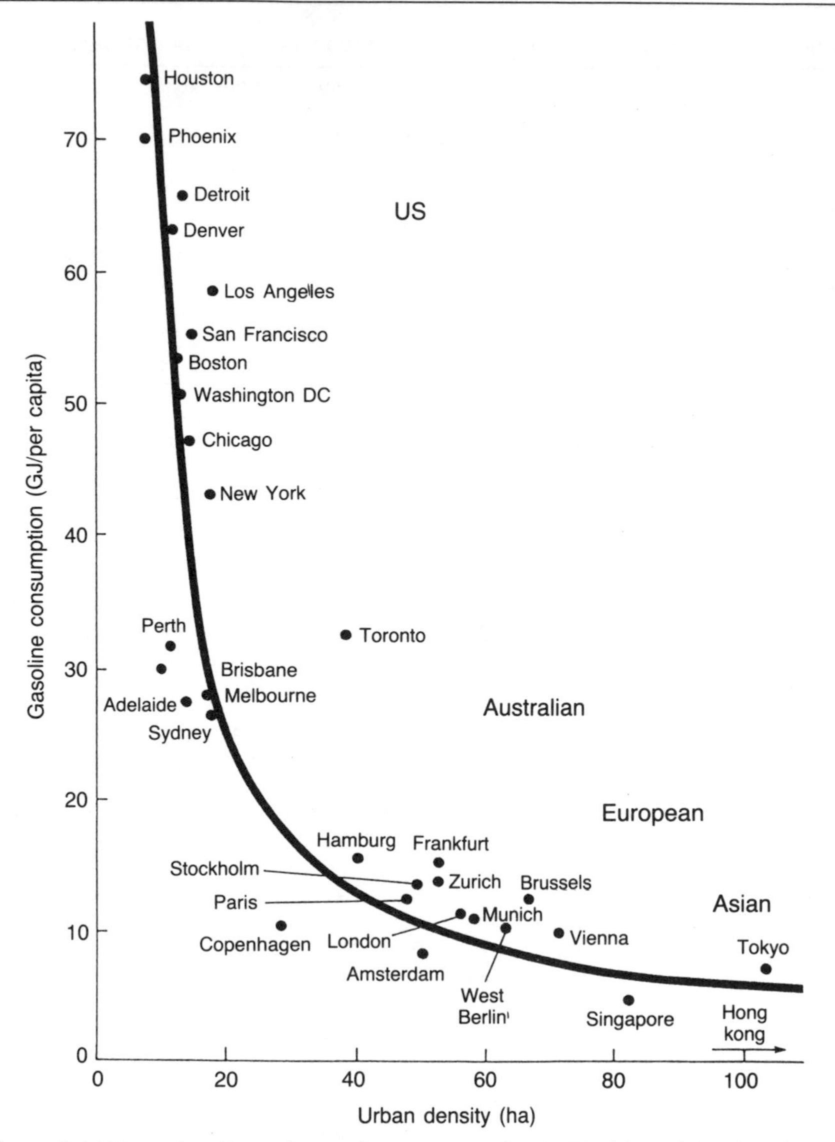

Figure 5.1 Urban density and gasoline consumption in 31 cities of the developed world in 1980. (Source: Newman and Kenworthy, 1988.)

dramatically within cities as between them. For example, in Table 5.4, the variation of gasoline use is shown across the generally car-based cities of San Francisco and Sydney. Transportation patterns are clearly linked to density in a similar way. Other data from Melbourne show the same kind of link between density and transit use (Kenworthy and Newman, 1991). Higher density, more centrally located suburbs are more

Table 5.4 Gasoline use per capita and density by suburb, San Francisco (1979) and Sydney (1981)

	San Francisco		Sydney	
	Gasoline (l/capita)	Density (people/ha)	Gasoline (l/capita)	Density (people/ha)
Central city	503	128	281	44
Inner suburbs	961	107	478	42
Middle suburbs	1 313	25	681	27
Outer suburbs	1 934	5	797	16

Source: Holtzclaw (1990); Newman, Kenworthy and Lyons (1990).

like European cities in their transportation patterns, while automobile-dominated outer suburbs consume some three or four times as much transportation fuel.

The data in Figure 5.1 appear to indicate a transition in urban design based on a transit to a car-based city when settlement densities reach somewhere between 20 and 30 people/hectare. The exponential increase in use as density falls, especially below 30 people/ha, should be a powerful guide for urban planners in a greenhouse future.

Also indicated is the difficulty of trying to reduce CO_2 emissions, fuel use and smog emissions through technological means if suburbs continue to be built at low densities on the urban fringe, while populations continue to decline in the older inner city areas.

Rebuilding our cities must reverse this process and concentrate land use, as well as providing different technological means of movement.

5.5 Economic and technological factors

Transport professionals discussing ways to reduce fuel and emissions have generally concentrated their policy discussions on economic factors, such as income and fuel prices, or on technological factors (i.e. vehicle efficiency). Undoubtedly these factors are important but our evidence would suggest they have been overemphasized. Consider for example:

- When our data on transport fuel-use by city was adjusted to show the effect of US fuel prices, US incomes and US vehicle efficiencies, the differences were reduced, but not substantially (Newman and Kenworthy, 1988). Well over half the variation between US and European cities cannot be explained by these factors. This is even more evident when US and Asian cities are compared.
- When the amount of fuel use per capita, by suburb, in Perth, Sydney and Melbourne was analysed, there was no correlation with income (Newman, Kenworthy and Lyons, 1990). Indeed, there is an accelerating trend in Australian cities for the outer suburbs to be areas of lowest income yet have the highest automobile use.

Table 5.5 Growth in per capita fuel use and changes in transit 1960–1980 in world cities by region

Cities by region	Additional fuel per capita 1960–1980 (l)	Change in per capita use of transit 1960–1980 (%)
US cities	584	–6
Australian cities	367	–39
European cities	226	+5
Asian cities	137	+32

Source: Newman, Kenworthy and Lyons (1990).

These two analyses again suggest that urban design is the fundamental framework in determining transport patterns.

The folly of attempting to reduce greenhouse gases, fuel and smog emissions purely by technological tinkering with cars is evident in the data on trends in urban fuel use and transit (Table 5.5). Despite technological advances worldwide in the period from 1960 to 1980 there continued to be a per capita growth in fuel use. Furthermore, it was greater in those cities that already had the largest automobile dependence, i.e. US and Australian cities. These cities also continued to decline in transit provisions, especially the Australian cities. By contrast, transit use in European and Asian cities continued to grow from an already very high level. It is worth noting that, in this period, Toronto's use of transit grew 48% per capita. The Canadians, as noted below, have done more to restructure their cities than most other automobile-based societies.

It is no wonder that Los Angeles still has smog more than 30 years after the battle to reduce it first began – the battle has concentrated on making vehicles cleaner and more efficient, but has done nothing to make them less needed. The notion often forgotten in this regard is the Jevons principle. The Jevons principle was first enunciated in 1865 when an assessment was made that improving UK coal-burning efficiency would save coal. Jevons predicted that it would in fact lead to greater coal use as the efficiencies would lead to more economic uses for coal. The same principle seems to apply to our present assessment of transportation fuel use. The price mechanism and urban sprawl ensure that for every increase in technological efficiency, there is a rapid increase in the use of vehicles.

5.6 Urban design solutions

It is not possible to reshape a city by simply redrawing town plans. Transportation priorities are the central force shaping urban design and hence there must be integrated design and transportation solutions. In our analysis there are three important practical integrated solutions – light rail, traffic calming and urban villages – which are central to any urban rebuilding process that can have an impact on greenhouse gas emissions. The solutions must also be socially and economically attractive. The transit-oriented urban design package we have developed has been applied in studies on two Australian cities – Canberra and Melbourne – and incorporates environmental, social, and economic reforms as integral features (Kenworthy and Newman, 1991).

The move towards this transit-oriented design is a central feature of design trends in US cities as they envisage a future constrained by fossil fuels. In a US Department of Transportation report (Rabbinowitz *et al.*, 1991a), transit-oriented development (TOD) and transit-corridor development (TCD) are outlined and guidelines provided. We have developed similar principles (see below) as a combination of solutions that could address the kind of technology shift and social organization that appears to be needed for 21st century cities.

5.6.1 Light rail

Rail systems have provided the chief competition for the automobile in the past, particularly in large, dense, old cities. Only rail could provide a fast, convenient service at a reasonable cost when old congested streets and city centres became choked with traffic. But such a service need not, and indeed could not, be provided realistically in modern cities, where the private car and bus were seen to be more appropriate for the less dense, less centralized urban form. This is how the argument has been presented for the past 50 years.

However, by the 1990s these automobile-cities have also become choked with traffic; they suffer from smog and noise, and vociferous objections from residents often block road authorities from building large new roads. Added to this there are deep-seated oil-supply concerns and greenhouse issues adding yet other dimensions.

Into this breach has come the idea of light rail – an electric, modern, tram that can run fast in convoy down main trunk lines and also has the flexibility to turn at right angles in urban centres and inner-city subcentres, and can run quietly and safely through pedestrian malls, or even immediately adjacent to houses. The mixture of environmental friendliness and low cost has meant that hundreds of cities, both large and small, have joined the light rail revolution in recent years (Bayliss, 1989; Taplin, 1990).

Most recent technological advances in light rail have consisted of:

1. the lowering of floors to enable street-level entry, thereby eliminating the need for platforms;

2. a capacity to run on heavy rail freight lines using mainline AC power and to switch to local light rail lines using DC power. This is being put into practice in Karlsruhe, where Wyse (1990) concludes:

> After building-up an effective and modern urban tramway system, within a few years Karlsruhe and its surrounding region will also have a completely integrated rail passenger network, achieved at relatively low cost and without massive civil engineering work or disruption. Light rail has been chosen in preference to buses because of its capacity, quality, environmental friendliness and lower operating costs per passenger. The city shows how the modal split can be influenced in favour of public transport without resorting to draconian car restriction policies or road pricing.

What is most significant about light rail is its attractive technology. It also seems to fit the kind of structural reform that automobile cities are needing. In most cases where cities introduce light rail (or heavy rail) an immediate process of urban concentration begins. This takes the sting out of the forces which disperse and sprawl our cities causing so much in-built transportation energy use and greenhouse gas production. In almost every case there is a flow-on of economic activity from the concentration of human activity. Light rail is also, therefore, of considerable interest to private investment.

Canadians have done most to rebuild their cities using light rail and to employ so-called 'value capture' techniques. Most new US systems, including the new Los Angeles rail network, also have a substantial private investment component.

The importance of light rail is that it offers a technology to modern automobile-dependent cities which can be introduced to low density/ dispersed areas and which can act as the catalyst in the generation of centres and subcentres.

The English-speaking world has been generally slow to back light rail systems. A study by TRRL investigated some of the reasons why the UK had missed the light rail revolution; the conclusion was that professional fashion, not economics, had been a major factor and that the overall urban benefit, including the civic pride it generated, had not been appreciated (Simpson, 1989). But recently the policy makers in the UK appear to be changing their attitude (Ridley, 1990):

> Transport studies carried out in Tyneside, London Docklands, Manchester, Sheffield, Bristol and Birmingham have shown that light rail is the basis for an effective way of dealing with their common problems of urban mobility and urban regeneration, and as 'green' issues assume increasing importance, a healthier environment.

5.6.2 Traffic calming

The other significant trend in transportation management has been the introduction of traffic calming techniques. These encompass a multitude of techniques designed to ease the impact of traffic on both residential streets and main roads, essentially by slowing it down. The idea has grown from the German 'Verkehrsberuhigung' and the Dutch 'Woonerf' (living space) which have the aim of slowing traffic and of making it easier and more attractive for pedestrians and cyclists. Traffic calming is thus the most significant policy orientation which facilitates walking and cycling (Newman, 1992).

Traffic calming is not just a set of engineering concepts (aggressive tree planting, particularly at the entrance to streets, different surface textures, raised levels, roundabouts, widened footpaths, angle parking . . .) it is also a community process. It is a way that a community can reclaim the streets and make them into more attractive public spaces.

Wherever traffic calming has been conducted on a large scale, it has been found that, contrary to many economists' predictions, the local economy has improved (TEST, 1989). This appears to be because:

1. people like to come to humanly attractive, green cities;
2. businesses like to locate in cities with a high quality urban environment;
3. car access is not banned, albeit not facilitated;
4. other modes of transport are generally facilitated.

This technique thus appears to be part of the practical and structural change which our cities require if they are to respond to the challenge of greenhouse, as well as the multitude of other issues associated with excessive automobile dependence. Traffic calming is the natural complement to rebuilding around rail nodes – both are needed if we are to fundamentally shift away from automobile dominance in our cities.

5.6.3 Urban villages

Assuming light rail and traffic calming are adopted, there remains a need to implement urban design concepts that both reflect the vision of reduced automobile use and provide the kind of human attractions and quality urban lifestyle desirable in a city environment. These are some of the new perspectives found today amongst city planners, particularly in Europe.

Recent examples of urban villages are: Arabella Park, Zamilla Park and Germering in Munich, Der Seepark in Freiburg and Kista in Stockholm. These are almost all private developments, and appear to be very popular with the residents. The characteristics of these urban villages are:

- mixed land use, with commercial offices and shops on main spines, surrounded by residential uses;
- high density, with everything within the 'village' within walking and cycling distance;
- extensive landscaping, including gardens on top of buildings and on balconies;
- a mixture of public and private housing, with an emphasis on family dwellings and thus quite large internal spaces;
- extensive provision for children, within easy view of dwellings;
- community facilities such as libraries, child care, aged centres and, in a few cases, small urban farms;
- pedestrian links with car-parks placed underground and traffic calming on any peripheral roads;
- public spaces with strong design features (water, street furniture, playgrounds, etc.);
- a large degree of self-sufficiency for the community, but with good rail links to the rest of the city.

Some successful urban villages have been built in Canadian cities (False Creek and New Westminster, Vancouver) and in Washington DC. In these cases the basic catalyst was usually the building of a new and popular rail system. In many cases, economists have been critical of the radical decisions to build rail facilities and suggested that incremental improvements to buses would have been sufficient in themselves.

Urban villages are the key urban design component which will enable the rebuilding of cities with reduced automobile dependence. They provide a lifestyle with minimal car dependence and the kind of densities which make rail highly viable, and biking and walking feasible due to the short distances between facilities and traffic-free areas. The evidence would suggest that those cities which have tried to build urban villages have found them to be an extremely attractive lifestyle option.

5.7 Conclusion

To minimize greenhouse gases from transportation requires a reduction in automobile dependence. It is suggested here that policies which incorporate a combination of new light rail technology, traffic calming and urban villages are required. This integrated solution to greenhouse emissions and automobile dependence in our cities is not a draconian measure designed to force people out of their cars, nor is it a return to some romantic state where a semi-rural lifestyle is advocated. The solution of light rail, traffic calming and urban villages involves available technology and appropriate land-use principles being merged and integrated.

We have quantitatively investigated some of the environmental and economic consequences of the integrated strategy outlined above. Several scenarios have been analysed, including one hypothetical case where all urban development in Australian cities was directed into urban villages and consequent modal shifts were accommodated by shifts in transportation spending. The results showed that in the five major Australian cities (McGlynn, Newman and Kenworthy, 1991):

1. 20% of per capita fuel use would be saved by 2005;
2. A$2 billion in infrastructure costs (power, water, sewerage . . .) would be saved by 2005, and A$5.5 billion by 2021;
3. 200 000 fewer cars per year would be needed by 2021;
4. new light rail and traffic calming to produce the required modal shifts could be provided entirely from a transfer of road funds no longer required;
5. A$0.6 billion per year in external cost savings would be achieved by 2005, due to reduce urban smog, traffic accidents and negative effects of noise pollution.

This integrated solution requires a commitment from both the public and private sectors and a recognition of their responsibility to find new ways of making our cities function without so much built-in car use. These are not only incremental steps; they are leading to phase changes in the way we manage our cities. The evidence suggests that those cities which are going fastest down this route are showing economic as well as environmental benefits. There is much to learn from the way in which Asian and European cities are designed and planned – indeed from any city structured around other than automobile-based assumptions and priorities.

If we cannot make this structural and attitudinal transition to rebuild our cities around reduced automobile dependence, and continue to emphasize incremental adjustments to motor vehicle use, we must also accept the likelihood that our children and grandchildren will have to cope with the consequences of our inaction – including the effects of global warming.

References

Bayliss, D. (1989) What's new in European and other international light rail transit projects, in *Light Rail Transit: New System Successes at Affordable Prices*, Transportation Research Board, Washington, DC.

Holtzclaw, J. (1990) Expanding urban density and transit impacts on auto use, CA Sierra Club, San Francisco.

Kenworthy, J. R. and Newman, P. W. G. (1991) *Moving Melbourne: A Public Transport Strategy for Inner Melbourne*, ISTP, Murdoch University, IMRA, Melbourne.

McGlynn, G., Newman, P. W. G. and Kenworthy, J. R. (1991) *Towards Better Cities: Reurbanisation and Transport Energy Scenarios*, Australia's Commission for the Future Ltd, Melbourne.

Newman, P. (1992) *Policies to Influence Urban Travel Demand*, OECD Group on Urban Affairs, Discussion Paper, OECD, Paris.

Newman, P. W. G. and Kenworthy, J. R. (1988) The transport energy tradeoff: fuel-efficient traffic versus fuel-efficient cities. *Transportation Research*, **22A**(3), 163–74.

Newman P. W. G. and Kenworthy, J. R. (1989) *Cities and Automobile Dependence: An International Sourcebook*, Gower, Aldershot, UK.

Newman, P. W. G., Kenworthy, J. R. and Lyons, T. J. (1990) *Transport Energy Conservation Policies for Australian Cities: Strategies for Reducing Automobile Dependence*, Murdoch University, Institute for Science and Technology Policy.

Newman, P. W. G. and Roberts, J. R. 'The effect of traffic calming in travel demand. *Transportation Quarterly* (in press).

Rabbinowitz, H., Beimborn, E., Mrotek, C., Yan Shuming and Crugliotta, P. (1991) The New Suburb. US Department of Transportation, DOT-T–91–12. Washington, DC.

Rabbinowitz, H., Beimborn, E., Mrotek, C., Yan Suming and Crugliotta, P. (1991a) Guidelines for Transit-Sensitive Suburban Land Use Design. US Department of Transportation, DOT-T–91–13, Washington, DC.

Ridley, J. (ed.) (1990) *Light Rail Review,* London Light Rail Transit Association.

Simpson, B. J. (1989) Urban Rail Transit: An Appraisal, Crowthorne, UK, TRRL, Contractor Report 140.

Taplin, M. (1990) in *Light Rail Review* (ed. J. Ridley), London Light Rail Transit Association.

TEST (1989) Quality Streets – How Traditional Urban Centres Benefit from Traffic Calming, TEST, London.

Wyse, (1990) in *Light Rail Review* (ed. J. Ridley), London Light Rail Transit Association.

Sustainable development, energy policy issues and greenhouse 6

Allan Rodger

6.1 Introduction

Is there not some irony in questioning sustainability of buildings and the built environment? Is there not clear evidence that these artifacts far outlive the cultural systems that bring them into being? Have not the Parthenon, the Roman Forum and the cities of the pre-Columbian America demonstrated that they have survived for many times longer than the cultures that built them and which they now represent to us? The simple answer to all of these questions is, of course, yes. Buildings and the built environment do survive, indeed, they persist. But in that characteristic lies one of their great advantages and a salutary warning. They often survive beyond their usefulness. In so doing they may act as a restraint on the changes that may be necessary for survival of the culture that they were created to support. Like a mill stone, they may be useful cultural artifacts but also, like a mill stone round the neck, they may herald and then contribute to the demise of a culture.

6.2 Reflections of the nature of capital

A distinguishing characteristic of our culture has been the way in which capital has been formed. We may well have been confusing ourselves

Global Warming and the Built Environment. Edited by Robert Samuels and Deo K. Prasad. Published in 1994 by E & FN Spon, London. ISBN 0 419 19210 7.

with words. For much of this century free market economies have been thought of as capitalist and the centrally controlled economies as not. Yet both have seen the formation of capital as of central importance in advancing their various objectives: increasing prosperity, increasing productivity, increasing world influence, etc. In this sense both are capitalist. The difference lies in who is controlling the capital.

The measure of success in all these capitalist systems has not, however, been in the simple accumulation of capital itself but rather in forming capital in ways that increased the capacity to produce and therefore to consume. Thus the capitalist societies have also been the producer societies and in equal measure the consumer societies.

The threat of an increased greenhouse effect embedded in an even wider environmental crisis challenges the very possibility of continuing increases in the level of consumption and therefore of production, and the inevitable environmental degradation that goes with these processes. Since capital has been, and remains, an essential and indeed limiting ingredient of the production/consumption system the environmental issues now faced impinge directly on the capital that is here now and that which we are currently equipped to produce. The challenge is to assess the relevance of the present capital that we have and to develop capital that would be appropriate in a sustainable future.

In terms of immediate gratification, forming capital represents a loss of the opportunity for consumption. In forming capital immediate benefit is forgone in favour of some presumed future flow of benefits and it can be done, on balance, only when these future benefits are valued ahead of, or greater than, the immediate benefit given up. It is therefore a matter of judging not only the quantity of the net benefits but how we and perhaps our successors will view these in the circumstance that prevail in the future.

It is essentially the process of making tools now for future use. Capital can usefully be viewed as a set of tools for the future. As with all tools it supports particular activities and is less well suited to others. Our capital therefore provides powerful cultural orientation, delimiting in advance our future options and our opportunities both for immediate use and for change.

Having concentrated for so long on the productivity of our economic system, relatively little is known about the capital basis of the production and consumption system. Recent studies by the Australian Bureau of Statistics (1990) shed some light on the matter. It appears that on the basis of their methods, definitions and assessments, some 75% of the accumulated capital stock is in the form of buildings or infrastructure. This is further subdivided: 30% to housing and 45% to non-housing and infrastructure (22.5% is in equipment while the remaining 2.5% is assigned to government charges).

There is limited information on the equivalent situation in other countries and there are, in any case, inherent difficulties in making comparisons from one place to another because of differences in the measuring techniques used. Even so, it is believed that the Australian pattern of distribution of capital may give a reasonable indication of what is to be expected elsewhere. It would be expected that Japan, with its high emphasis on capital equipment and its relatively low provision of relatively low performance space per person, to have less of its capital wealth in the built environment. On the other hand, it may be expected that many poor countries will have less of their wealth in the form of productive equipment than Australia.

This being the case, and capital having this key role as our tools for the future, the importance of the built environment as the dominant element in this kit of tools cannot be underestimated.

There is a sense in which tools are passive in that they become productive and interact with the environment only when they are used. This is a characteristic that seems to be under our control. A tool can be used or not – at our discretion. The built environment shares this characteristic with other tools as when building services are used (heating, lighting, cooling, lifts, automatic doors, etc.). It appears that there is a choice to draw down resources and produce environmental impacts or not. Yet the built environment also has the characteristic that there is little choice as to whether it is used or not. Such is its importance for survival that we cannot choose not to use it. There is a complex dependency relationship with buildings and the built environment that leaves very little discretion. Once a built environment is formed, there is little opportunity other than to use it. In making it, substantial commitments to future use, and therefore to the nature and extent of future environmental impacts, are made.

The reason that there is little option but to continue to use the built environment may appear obvious – where would we be if we were not settled in our existing built environment? But the issue is more complicated than this. Certainly shelter is required, but more important is the need for a full comprehensive and continuous life support system and that system is the built environment; it is the medium through which access is gained to high priority services. It is the water supply and waste disposal systems. It is an integral part of the food processing, storage and distribution systems, on which we have become totally dependent. It is the place where goods and services that can subsequently be exchanged for other goods and services are produced. Not least, it acts as a wealth distribution and redistribution system as we draw on each other for services and thus spread around the capacity to consume.

6.3 The thermodynamics of environmental change: which way is forward?

For any system in which energy is transformed from one form to another there is a maximum efficiency at which the system can operate. The maximum is determined by the limiting conditions of the system, such as the maximum permissible temperatures and pressures and the temperatures and pressures of the ambient environment. Carnot (1824) defined this for the idealized gas engine. The principle is universally applicable.

In practice, no machine reaches its theoretical maximum because the machine itself introduces inefficiencies additional to those of the basic thermodynamics. Thus, when a system is established and defined in terms of its processes and its limiting operating conditions its maximum possible efficiency has been established. The aim would then be to develop equipment that would allow the system to operate as closely as possible to this maximum. History suggests that the first efforts would fall far short of this target. A process of further research and refinement and redesign would then progressively allow improvements to be made. There would be two kinds of improvements. There would be improvements in efficiency through refinement and design (e.g. better seals, better lubricants, more efficient fluid flow patterns, etc.) and there would be other improvements through changing the limiting conditions (i.e. introducing new materials that allowed the system to operate at higher temperatures or pressures).

At each stage in this progression, however, it becomes progressively more difficult to secure a smaller and smaller improvement towards that theoretical maximum limit. Eventually it becomes infinitely difficult to obtain that final infinitesimal improvement. The benefit becomes hopelessly outweighed by the effort needed to secure it.

This is in the very nature of improvement through improving efficiency; progress can be achieved but it remains progress towards a limited objective and it becomes progressively more difficult as the ultimate limit is approached.

When the improvements that are needed are greater than those that can be achieved through improving efficiency, or when the effort involved is greater than the benefits that might become available, the approach to the problem through increasing efficiency cannot succeed. It is the system itself that must become the arena for exploration. It must be by choosing different systems and establishing new limiting conditions that the possibility of the required improvements can be established.

It was Edward de Bono who pointed out that no amount of effort put into digging one hole (i.e. efficiently doing the task that we have already defined) would ever result in the production of two holes (the establishment of a new objective)! For that to happen there must be a discontinuity. There must be a decision to dig another hole.

What, then, does this have to do with the environmental problems now faced and in particular how does it affect the approach to the problems which are associated with an enhanced greenhouse effect? For much of human history we have been engaged in harnessing energy for our purposes and improving the efficiency of energy transformation. It is through such processes that humans with their technology, artifacts and their harnessing of natural systems have grown into the natural environment. More recently humans have gained access to the accumulated energy stored in the fossil fuels, harnessed it with machines and greatly increased the capacity to transform materials and energy. Stephen Boyden of the Australian National University has engagingly called this techno-metabolism.

This process has inherently been one of growth. In the past there have always been new lands to enter, new resources to be tapped and the broader environment has usually had the capacity to absorb and deal with the by-products of this development. In general this process has continued, though there have been local instances of overloading and collapse of the ecosystem or the depletion of non-renewable resources or the build up of intolerable levels of pollution or environmental degradation.

The situation now faced is different. The cumulative effect of human actions now influences the whole of the world system. Environmental problems in any one area are no longer isolated from problems elsewhere; collapses anywhere are felt throughout the system. Collectively human activities are pressing on the limits of the biosphere and locally they frequently transgress these limits.

The greenhouse effect epitomizes this new situation. A tonne of carbon dioxide in the atmosphere is a tonne of carbon dioxide irrespective of where it was released and irrespective of who or what was responsible. The atmosphere acts as the great integrator and the evidence is that it is beginning also to act as the great limiter on some of our activities.

This poses the energy question in quite a new way. It foreshadows the day when we will have to operate within the capacity of the atmosphere to absorb our wastes, and in particular our greenhouse gases, while still providing humans and the rest of the biosphere with sustainable conditions. The question changes away from growth and efficiency to working within these new bounding conditions.

This is not, of course, to abandon or in any way denigrate the idea of efficiency. Efficiency in the transformation of energy will remain a key part of our armoury in our continuing struggle for survival. It will, however, have to be seen as a refinement of systems that are inherently compatible with overall environmental requirements. Attention will have to be redirected to designing systems that are inherently based on low levels of energy consumption and on efficient energy conversion technologies.

6.4 Ethics of sustainable development

The idea of sustainable development presents challenges far beyond those normally identified with the built environment. When it is viewed as a major component of our set of tools for the future the built environment must also be viewed as a significant participant in determining future levels of environmental impact and thus participating in decisions about the sustainability of the biosphere and the role of humans.

It is only once sustainability cannot be taken for granted and when it is clear that human action is the cause of this uncertainty that responsibility for the future emerges as an issue. It is then that the flows of material and energy through the biosphere become a question of ethics, and the questions become unavoidable. Responsible for the future of the biosphere, and for how long? Responsible for all species or the retention of a sustainable ecosystem? Responsible for ourselves, our grandchildren, how many generations, other people's families, the country, other countries? How far do our responsibilities reach over space and time?

At a more practical level, and even once we accept responsibility, we must also ask ourselves what we can actually do. How much needs to be done and how quickly? Not least we must ask ourselves whether we have the personal will and collective political commitment to do what we conclude needs to be done and what should be done.

6.5 Across the scales, across the professions

It has been convenient through our long period of materialist growth to simplify our problems by dividing them into sub-systems and components. Underpinning the industrial revolution with its single purpose technologies, has been Adam Smith's social dictum on 'the division of labour' (Smith, 1961). Professionalization has been part of that process. The economic and development advantages of specialization have been won at the cost of a reductionist approach to our problems.

Nowhere is this to be seen more clearly than in the approach to buildings and especially the complex interactive systems of the built environment. Perhaps this is to be expected because the habitat is an extremely complex array of social, economic and technological variables. It would be argued by many that it has been possible to achieve what has been achieved over recent times only through the division and specialization of labour and the consequential separation of the various elements. If that has resulted in there now being some problems, then perhaps that is to be expected and even accepted as a legitimate price to be paid for the advantages that have been secured.

Yet the conditions within which this 'divide and conquer' approach has been successful may now have changed so dramatically that the divisions established within the context of the growth and consumption culture are no longer appropriate to the new and evolving circumstances. When the question changes away from, 'How much are we able to produce?' to 'How can we develop sustainable lifestyles and support ourselves within the bounding limits of the local and global environment?' new criteria emerge. With them arrives the need for development of expertise and action across previous divisions.

It must not be assumed, however, that there can be a new order without specialization. These are continuing systems which are essential for survival, at least in the short to medium term, and which are fundamentally based on those divisions and specializations that it itself fostered. What should be looked towards is some reappraisal of these divisions and the evolution of new associations, new structuring of knowledge and the development of new technologies. The lesson that must be learned is that social and technological change are now essential elements of any strategy for sustainability. Thus any new structuring of the division of labour should be seen as important but it should also be kept in mind that in this continuing period of uncertainty and change new arrangements should not be expected to be more persistent than those that they replace; they too must be expected to change and continue to evolve with the problem.

The built environment professions are slow (and understandably so) to respond to the fundamental changes that are occurring in the social and physical environment within which they operate. They are still structured in much the way that they themselves arranged through the establishment of institutes and systems of registration over the past 150 years. Yet the problems now confronting society and the professions is remarkably different. The inefficiencies that were an acceptable *quid pro quo* in the era when growth in the consumption of material and energy was a principal indicator of progress, are now major obstacles on any route towards an equitable and sustainable future.

The provision of water may provide a good illustration. Not so long ago it was the mission of water engineers to provide as much water as could be used at the lowest possible price. There was little detailed consideration of the idea of need. All water was to be provided at the highest standards of cleanliness. There was little consideration of the upstream environmental/ecological effects of dams and catchments nor were the downstream effects of reduced stream flow or eutrification and pollution or other forms of environmental degradation seen as fundamentally bad. Where problems were identified this was seen as an inevitable and acceptable part of progress. One of the attractions of the approach was that it was largely in the hands of a relatively small number of experts.

Now, however, it is recognized that an environmentally friendly water supply system sets out to achieve quite different objectives with different technologies, with different equipment and embedded in fundamentally different decisions making processes. Within the new image there is an expectation of very careful and selective use of water with good provision for economizing. It would be expected that water should be reused, wherever possible. It would not be expected that all water should be at a potable standard, but it would be expected that toxins were excluded from the systems and that nutrients were removed or withdrawn, and that water eventually returned to the natural system would be done with minimum disturbance to the downstream ecosystem. In this scenario detailed management of the water stream in the household, factory or landscape (gardens, market gardens, parks, agriculture, etc.) would be an essential part of the overall conserver ethic shaping the system as a whole. Its creation and use become a cooperative and collective activity with important and well informed decisions being made at all levels in the system.

Such a scenario is one of the primary generators for the development and design of a sustainable built environment system. While at first it may seem to be remote from the issues of the greenhouse effect this is more apparent than real because the establishment of the conserver ethic together with the equipment to support a society basing its lifestyles on relatively low flows of material and energy is the essential ingredient for sustainability. The distribution of knowledge, access to knowledge and decision making in such a network of active and interacting participants is quite different from the centralized and hierarchical systems which have been developed in the past, (in which, ironically, knowledge and expertise were considered to be scarce). The relevant professional contribution in this new situation is inherently multi-disciplinary and of necessity operates in an educational, participatory and facilitating mode.

How then can these ideas be applied in exploring possibilities for sustainability in the greenhouse era?

6.6 Strategies

It is said that there are three strategies being advanced for dealing with the greenhouse effect:

- ignore it and hope it will go away – the no action strategy;
- accept that it will happen and prepare to deal with the consequences – the coping strategy;

- accept the scenario of impending climate change as anthropogenic and undesirable and take action to prevent or limit its progress – the limitation strategy.

There is a long history of ignoring threats and sometimes, of course, of perceiving threats that have been unfounded. The greenhouse effect is difficult to ignore since we already owe our existence to a natural greenhouse effect that has been essential for the whole evolution of the biosphere as a living system. It is also known that human activity is causing change at an unprecedented rate in the driving forces that create this greenhouse effect.

In some countries, particularly Australia and Sweden, there are arguments advanced that the greatest threat to the continuation of the growth culture (and its hierarchical political, commercial and knowledge structures) is to undermine confidence in the continuation of that culture. This argument holds that the technological advances that would come from pursuing the growth culture with confidence and vigour would surely provide solutions to any of the (minor) problems that it might produce. In this view a future, but as yet unknown, 'technological fix' (or fixes) would overcome the problem. Refusing to allow present perceptions of problems to interfere with progress is therefore a necessary part of such a strategy. It is a strategy closely associated with the writings of Herman Khan* and well described by Robertson (1978) as the Hyper Expansionist (HE) strategy as he then proceeds to advance his alternative Sane Human and Ecological scenario (SHE).

More specifically this strategy would argue that continuing to release greenhouse gases in support of a rapidly developing technological culture would itself result in the development of knowledge and technological capacities that could then be harnessed to resolve any problems that might arise or have arisen. It is a call for a great act of faith in some 'hidden hand' steering the destiny of the world (Smith, 1961).

For whatever reason some people and some countries may support 'no action' strategies it now seems inevitable that the greenhouse effect will be enhanced even on the basis of greenhouse gases that have already been released and further releases to which we are already committed. It is therefore inevitable that there is a need for a coping (or adaptation or palliative) strategy to deal with the effects as they emerge and to take defensive action in advance of atmospheric and climate change.

All interventionist strategies require some assessment of the likelihood

*Herman Khan's writings epitomize the ideas of those who support the appropriateness and inevitability of human supremacy over nature 'as the great transition' from a state of poverty and powerlessness before the forces of nature to a state of affluence and control over nature. He also points out that success in this strategy will marshal its own support as 'more and more nations will have a stake in following this transition through to its logical conclusion'.

of the intervention being successful. For coping and limitation strategies the criteria on which action might be planned are, however, quite different. When planning to cope it must be known what it is that is being dealt with; the probability of specific effects in specific places must be known and plans of actions that would be effective in defence against these changes must be developed. If, on the other hand, progress of the greenhouse effect is to be limited, actions which would achieve that objective need to be determined. Essentially ways of limiting, or preferably reducing, the net release of greenhouse gases must be devised. The two sets of actions, each appropriate to its own objectives, are likely to have little if anything in common.

Since buildings and the built environment are fixed in their location any planned adjustments must relate directly to changes that are anticipated locally. Action on a coping strategy is therefore dependent upon reliable knowledge of the detailed local changes that are likely to occur, the rate at which they will occur and the stage at which action would be justified.

In sharp contrast, action to reduce the progress of the greenhouse effect can be almost independent of any of the local implications of the greenhouse effect and climate change. Unfortunately in some ways, perhaps fortunately in others, knowledge of the greenhouse effect is still rather unreliable at the local level and more reliable at the macro/global level. The basis on which to make plans to cope with the anticipated greenhouse effect changes is therefore much less reliable than that on which action for a limitation strategy could be planned.

In formulating appropriate action within a coping strategy it is necessary to assess the confidence level associated with each of the predictions for local change. The most locally predictable and dramatic result of an enhanced greenhouse effect is likely to be changes in sea-level. This is not because the projected changes are large but rather because even very small changes are likely to have spectacular local effects with far reaching global implications. A quite disproportionate share of the world's population lives near to sea-level. Some of them, particularly in river deltas and some of the island chains in the Pacific and Indian Oceans, will lose their habitat. In some cases the predictions are that their whole national territory will be lost with even small rises.

There are also many cases where the sea-level effects will be greatly exacerbated by changes in storm patterns. It is anticipated that alongside the rises in temperature there will be a displacement of climate patterns towards the poles – it has been estimated that for a 3°C rise in the average global temperature there would be a rise of 10°C at the poles and 1.5°C at the equator, resulting (typically) in a poleward migration of climate patterns and correspondingly of plants and ecosystems. This will include a change in the location of the tropical cyclone belts. Such combinations of effects can be expected to produce rises in maximum (flood) sea-level of several metres. In resort areas, for example, where the main purpose of the settlement may be to gain close access to the sea, such a change can be expected to have dramatic effects on the pattern of development and cause major disruption in existing settlements.

An important, but often overlooked, aspect of sea-level rise will be changes in groundwater regimes. In particular, ingress of salt water up estuaries can be expected to affect the productive capacity of coastal lands which are often the home of the poorest section of Third World communities. Contamination of those fragile fresh water lenses on which many island communities depend for their water supply, and therefore their survival, can also be expected.

These effects will limit opportunities for development and will damage or destroy existing habitat. They will contribute to local population problems and then create environmental refugees. It has been estimated that up to 100 million people will be displaced. The great majority of these will be from the island communities, from the delta areas of the Ganges/ Brahmaputra and the Nile and from other low lying coastal areas in Thailand, Cambodia, and of course India and Bangladesh. Others will join this growing stream of refugees as rising temperatures and changing climate patterns inexorably cause ecosystems to migrate and cropping systems to be displaced. These will be the economic refugees who can no longer support their way of life as their environmental support systems move away or are destroyed by climate change.

Alongside these movements of people and ecosystems national boundaries can be expected to remain static, thus precipitating complex political pressures. Total loss of habitat is clearly the most severe type of impact on our buildings and built environment. It presents us collectively with the need for complete replacement of a whole life support system with all its infrastructure.

Though less dramatic than complete loss through inundation, or destruction of human habitat, or migration of ecosystems, the other impacts on the built environment are likely to be much more extensive and affect vastly more people.

Rising temperatures will create need for changed levels of insulation, different patterns of energy production and consumption, new codes for design, different kinds of building equipment and services, increased need for outdoor shade and in higher latitudes better opportunities for the use of solar energy, but also higher rates of corrosion. Some, but not all, of the effects of rising temperatures will be bad. It is difficult in Scotland and Tasmania, for example, to generate much enthusiasm for action that would work against an increase in temperatures!

This somewhat jocular remark masks two profoundly important aspects of the anticipated enhanced greenhouse effect. The changes envisaged are expected to take place at about fifty times the most rapid rise in temperature experienced in the emergence from the ice age about 150 000 years ago. Since the problems faced by ecosystems in adjusting to change is profoundly related to the rate of change, the presently anticipated rates of change present an ecologically disruptive and environmentally damaging prospect. The other aspect that the remark reveals is that despite the seriousness of the overall situation different people will have quite different experiences and will be expected to hold different views on what they are prepared to do and what should be done. The very nature of climate change and the effects that are likely to flow from it will make it difficult to assemble the coherent political will that the situation needs.

Changed rainfall patterns and corresponding changes in run-off and ground water regimes will present the need for changes to existing buildings and urban systems. These will range from foundations that shrink (drier) or swell (wetter) to gutters that are no longer adequate, seals that leak and overhangs that no longer give adequate protection. With changes in actual rainfall will come changes in humidity and consequential changes in biological regimes. Areas requiring protection from flies and other insects will change and buildings themselves will be subjected to new patterns of attack by new distributions of animals, algae, fungi and borers.

It is not so much that the new requirements will be worse than the old ones that caused the problems. It is that any change in the climate and the biological environment presents a problem. In this sense the built environment is inevitably a victim of change. Designed or evolved for one set of circumstances it will be less well suited to almost any significant environmental change, and so a process of development must ensue as the built environment is readjusted to suit the new and evolving environment with which it will have to contend.

The built environment is not, however, a hapless and innocent victim. There is an important sense in which it is the principal villain actively participating in promoting the greenhouse effect. A simple analysis of energy use shows that energy is, in varying amounts, used to heat, cool, light and service buildings. Almost all of this derives originally from burning fossil fuels and this implicates the release of carbon dioxide. Energy is also used through the construction industries to make basic building materials (steel, aluminium, glass, cement, bricks, plaster, etc.). These are major energy consuming industries and consequently they are major producers of greenhouse gases. Still more energy is then used to shape and fabricate natural materials such as wood and earth and still more in transport and the assembly of construction materials and components into the final building.

As with individual buildings an urban system is not useful until it is supplied with energy to make it work. Thus, energy used in transport must be seen as an essential part of the built environment energy budget. It is only when a building is connected with other buildings through a transport system, that it functions as part of an effective and comprehensive life support system.

So accustomed are we to dividing up our world (and our budgets) that we find it strange to consider transport and transport energy as part of the built environment. A moment's reflection on the role of vertical transport by lift, however, may clarify the matter. A lift is clearly part of a tower block and the energy it uses is part of the energy budget of the building. And so at the urban scale mechanized transport systems and the energy that they require are essential components of our urban life supporting systems. The upper parts of a tower block are useless and therefore valueless except when provided with vertical access: both the lift mechanism and the continuing supply of energy to drive it. In the same way the suburban house can be useful as part of an effective life supporting system only when it is supported by reliable horizontal access systems, typically in the form of private motor cars and reliable fuel supplies.

Further accounting anomalies tend to obscure the overall significance of the built environment as a primary producer of carbon dioxide through its energy needs. Where does steel go if it does not go into buildings or transport? And, similarly, for many other materials and products of manufacturing industry. Much of it is eventually attributable directly or indirectly to the creation and running of the built environment.

The built environment also has a propensity to create and then perpetuate a pattern of energy usage. It has been calculated that the electricity used in the United States for drying clothes is approximately equal to the total amount of energy produced by their nuclear power programme. This amount of energy is now needed because of the development of certain lifestyle patterns reinforced by community expectations (for not hanging out washing), regulations (that prevent hanging out washing), changed community expectations (that more women are away from the home during the day and security is not good), and by the fact that many people do not have the space and most people do not now have the equipment to dry clothes in the wind. At every stage it has been convenient but the cumulative effect of myriad little decisions about the form of the built environment has created dependence on this energy to support our home-based clothes washing activities. In creating one system an alternative has been dissembled. Having made choices they have been supported with space and equipment, and the commitments for energy, the release of greenhouse gases and other environmental impacts far into the future have been made.

When this realization of how the present built environment commits us to future patterns of environmental degradation is combined with the fact that the built environment, even without much of its rolling stock and vehicle fleets, accounts for the bulk of our formed capital, the evidence is compelling that there can be no resolution of the greenhouse effect, and indeed other major environmental issues, without some major contribution from this dominant cause of the problem. Such is the scale and involvement of the built environment that the victim and the villain must become the 'white knight' (Rodger and Robertson, 1991).

The principal task in establishing a strategy for sustainability in the face of the greenhouse effect must be to describe built environment systems that would be sustainable and would support sustainable lifestyles. But an open range of choice is not available. Various constraints must be met. First and foremost is that there are already built environment systems on which we are wholly dependent for survival and which provide the only basis that we have for working through various social and environmental problems to a sustainable future. The old proverb that every journey starts with the first step might be extended to remind us that the first step must be from where we are. Where we start from is already fixed. Our choice is in the direction in which we move and the rate at which we proceed.

Possible futures are strongly conditioned by these factors and argument rages about both of them. Let us review them to see what can be learned, starting with the observation that we must start from where we are. Where else?, might be the obvious response, but it should be recognized that much of the work that has been done on possible future built environments deals with new towns and new constructions on clear building sites. Throughout history there have been utopian propositions about appropriate urban forms and new towns (see Ebenezer Howard, Le Corbusier, Paolo Soteri, etc.). Typically they choose to start with a clear site, or are happy to clear away part of an existing city to make a clear site. While these have been useful, and in some cases influential, contributions to thought, they generally fall far short of a realistic strategy for the future. When the focus of attention is on the physical form of the proposed new environments and with new social structures, as it often has been with architects and planners, this is a logical approach and it can be an appropriate approach when the proposals have to deal with only some very small part of the total built environment and a correspondingly small part of the population. The situation with which we are dealing in the case of the greenhouse effect, however, involves all of the built environment. Since collectively another built environment cannot be constructed alongside the present one and then occupied (as may be done in the case of a single building) what is done is limited by the existence of present structures and services, and we must set out to convert, supplement, and where appropriate extend this system to our new purposes.

This is not to argue that there is no place for completely new settlements or extensions to existing settlements that are significantly different from what went before. It is to assert, even to insist, that the principal task before us is the redevelopment of the existing built environment. Where there are to be new constructions these should be seen at each stage as extensions and additions to the existing structure. As such they will be greatly influenced by the other urban structures to which they will have to relate.

The nature of the relationships that may be developed are many and varied. For example new technologies may be introduced to supplement and perhaps ultimately replace an old decaying infrastructure. New materials and construction methods may be created alongside existing systems still needed for the existing stock. In some cases new patterns may be developed where that can be done without interference or in ways that are complementary to existing situations.

This approach has far reaching implication for the options that should be investigated and for the methods that can be employed in developing designs and propositions for action. Firstly, there should be consideration of what can be done to make best use of the existing urban fabric with the resources already available. Within this, ways to use these resources in ways that will reduce the emission of greenhouse gases should be sought.

6.7 White knight scenarios for sustainable development

6.7.1 Salvation by densification

Undoubtedly the most widely discussed, and fashionable, approach to urban development and redevelopment is characterized by the popular catch phrases, 'urban consolidation', 'densification', 'medium density infill housing', all of which are to be associated with 'efficient public transport', 'pedestrianization' and 'bicycle tracks'. Together these support the idea of dense urban areas with efficient internal public transport. The work of Peter Newman and his colleagues (Newman, Kenworthy and Lyons, 1990) has become closely associated with this approach in Australia and around the world. It is also an approach widely espoused by the architecture and planning professions.

Drawing on historical precedent, this scenario for development advocates that urban development should concentrate on those urban forms, current examples of which appear to demonstrate low levels of energy consumption. Thus, having found that existing cities with high population density and good public transport systems typically now operate with less internal energy per capita than diffuse, low density cities based on private motor transport, the advocates of this scenario would seek to modify our present urban forms by developing higher density in support of public transport. This is design for a future based on experience drawn from the past. It has a persuasive logic and there is no reason to doubt that it can be applied to good effect in some circumstances.

Such a strategy, however, should be carefully assessed and, if applied, it should be applied with caution. Perhaps our first observation should be that densification seems to be a change in exactly the opposite direction from much recent development. One must therefore ask why there seems to be such a strong demand for lower densities and suburbanization and whether this demand will persist.

As distinct from the theory that may indicate benefits to be gained on the basis of certain limited criteria (such as greenhouse gas emissions or depletion of non-renewable resources), practice demands that the reality of community aspirations and expectations, as expressed through decisions being made in the market, must be taken into account. Current development is actually at relatively low density and the community, through its purchases and other choices, clearly supports that form of urban development. Any densification strategy must therefore present the community with options that are attractive enough to draw home-owners, buyers and investors (private and public) away from low density suburban development and into higher density urban environments.

The advocates recognize this and logically advocate a variety of measures to counteract it. Community education extolling the virtues of urban living is to be supported by regulation, financial incentives and government initiatives. Even so there remains the worrying perception that current proposals for densification smack for many people of 'putting the clock back' and returning to a less convivial pattern of living. It may reasonably be expected that once the idea of suburban living has been rampant for half a century or more the images and expectations that it has developed will not easily be discarded. Once the genie has escaped from the bottle it is hard to put it back in. Our commitment to diffuse low density suburbia served by private motor transport (and almost impossible to service by public transport) may already have gone beyond a point of no return.

Quite apart from this set of cultural aspirations there is the question of growth. It would now be widely accepted that growth in global population is, in the long run (but preferably in the short run) unsupportable. Densification, as it is currently being portrayed, is inherently a strategy based on an assumption that there will be continuing growth. In this sense it is a strategy for supporting a pattern that is itself undesirable and must therefore be called in question.

Particularly since the densification scenario derives its support largely from historical examples of the part it may have played in establishing current high levels of dependence on energy, some assessment of these processes is worthy of further consideration. If we take a longer time scale than Newman and his colleagues and look at the way that our urban systems were formed and are still being formed it can be seen that urbanization itself has a close relationship with increasing mechanization and with increasing levels of energy use and dependence. Modern cities grow partly from within their own populations but also by inward migration from rural areas. Urban settlements, with or without efficient public transport, are high energy life support systems when compared with the energy normally associated with traditional rural living. However, a distinction between energy and overall environmental impact must be drawn – some low energy rural lifestyles are damaging to the environment, most likely when the population is under great economic stress. A more technical issue is whether efficient public transport can actually achieve results that are compatible with what may ultimately be required. Viewed globally the greenhouse effect is likely to set limits to the release of greenhouse gases that cannot be met for an expanding population merely by making transport systems efficient. Densification is a strategy which does reduce the demand for transport around its transport nodes by virtue of the increased density. It then seeks to support transport between nodes by efficient technology. The question is not about efficiency but rather about cumulative environmental impact.

Public transport is often lauded as being environmentally friendly. This too needs careful assessment. While it is true that a full commuter train can move people at a much lower level of energy expenditure per passenger kilometre than a private car it is also true that most trains run nearly empty and that occupancy rates are as low as or lower than cars. This is compounded by the fact that efficient public transport tends in time to increase commuter distances as speeds increase. The effect of this is that in the final analysis there may be little overall difference between the two. There is some evidence from studies in Melbourne that motor transport in that city is more energy efficient and more efficient in terms of greenhouse gases that a centralized rail system would be.

Perhaps the other major area for concern is the limited range of life supporting systems that are included within this historicist approach. The goods and services drawn from urban hinterlands are taken for granted. Water supply and waste disposal have been largely external to our urban systems. Ambient and renewable energy (solar, wind, hydro-electric, geo-thermal, methane, alcohol, fuel wood, etc.) have been peripheral issues remote from our centre of interest and attention. Most importantly, food is produced almost exclusively outside our living areas yet the physical relationship between rural production and urban consumption, with all the processing, packaging, storage and distribution that this necessitates, has resulted in an agro-urban system that is extraordinarily dependent on fossil fuels and the continuing release of greenhouse gases. The question that now hovers menacingly over all habitat development is whether this separation of functions, and the environmental and resource implications that go with it, can continue to be supported. The question extends further, of course, than our existing settlements because increasing population and the increasing need for shelter that goes with it are being accompanied by increasing levels of urbanization within the agro-urban system. The pressing strategic question that may ultimately determine the longer term well-being of humankind, and even the biosphere itself, may be how the human habitat is shaped in support of this growing population. The environmental impact, in particular the contribution to the greenhouse effect and climate change that is implicit in the form of such development, may well be the determining factor. Given the overriding importance of water, food, shelter and energy, in any viable life support system, the provision of all of these on a sustainable basis becomes an essential criterion against which urban strategy must be judged.

There are also complex issues of the interaction between existing urban systems and the way in which they are extended. Not least there are issues of image and conceptual continuity. These are powerful influences operating not only from one place to another but also across cultures and increasingly across communities at different levels of technological development. There is no doubt, for example, that the patterns of development pursued by the rich are often copied by the poor. This is driven by images of the desirable lifestyle. Through such social and psychological processes development patterns are interconnected. Inevitably the actions of those who are seen to be leaders, be they individuals or technologically developed nations, influence all those who follow.

These processes put great responsibilities on those who, for whatever reason, are deemed to lead. Once this set of inter-relationships is recognized and understood it can be seen that sustainability is not something that can be achieved by having the poor develop in ways different from the rich. Rather the assumption of the rich must be that the poor will follow. In time they will try to have the same patterns of environmental impact through attempts to increase their level of consumption: the so-called standard of living. It may even be that, without the advantages of modern technologies, the environmental impacts of their efforts will be greater and for less advantage than those of the rich. Through these mechanisms the leaders of the world community must now consider not only the effect of their own actions and assess whether it is environmentally acceptable but also whether the proliferation of such actions by vast numbers of others would be compatible with a sustainable system of settlement. They might well ask themselves where lies the best prospect for themselves and their successors. To lead humanity, if the route takes us to a green and pleasant land flowing with milk and honey and with plenty for all, may indeed be an attractive role. If, however, the rich developed world is leading humanity towards an environmental precipice they may discover too late that the momentum of the madding crowd behind will sweep them and many others to catastrophe.

Having raised these cautionary issues it is appropriate to return to the matter of densification. Despite these various caveats there do appear to be ecological and environmental niches within which such a strategy would be appropriate. Principal among these are situations in which existing infrastructure facilities are under-used and where medium density infill can draw on this spare capacity while also enhancing the efficiency of existing public transport. Some unused or under-used inner urban land, frequently already in public ownership or becoming derelict as industry moves outwards, presents such opportunities.

6.7.2 Sustainable suburbia

Quite a different approach to reducing the impact of urban systems on the broader environment and of reducing the contribution to the greenhouse effect is to start by considering the habitat as the life support system for the human community. The logic of this approach is to assign priorities and to assess how the products and services necessary, first for survival and then progressively for a congenial lifestyle, can be delivered in ways that are compatible with the integrity of the biosphere. It must immediately be admitted that to carry these admirable and perhaps altruistic intentions through to their logical conclusion, even at the theoretical level, is a daunting task and one that has not yet been done in anything other than a cursory manner. Even so, it provides an approach to the development and critical appraisal of alternative strategies.

The question becomes how to contrive a low impact life support system and then how to modify our existing systems towards that objective. Throughout all of this, and to motivate the transformation process, it is necessary that the results at every stage are attractive and, ideally, convivial.

As before, the movement of people and goods must be a central generating factor. This movement is the method of re-balancing the relationship between the supply and the demand for people and goods. It is a crude and environmentally degrading process consuming resources and human time through an otherwise unfulfilling activity. There are therefore compelling environmental and social reasons for trying to develop a low movement society through the creation of locally complementary sets of life supporting facilities all within reasonable walking or cycling distance of each other. The degree to which these facilities form a comprehensive life support system will be a measure of how effective the proposed system would be in achieving its environmental and social objectives.

To interpret this idea in more specific terms and in the context of a real built environment it is instructive to assess the extent to which a specific urban area meets the criterion of comprehensiveness as a life support system. Take a single site, a hectare, 10 hectares, 100 hectares or a small town. The limit for the area that it is appropriate to consider ranges up to that at which pedestrian and bicycle access is no longer practical for a significant proportion of the population and for a significant number of the journeys that would be needed. The results are rather surprising and reveal the extent to which our present urban systems fail to meet what appear to be reasonable environmental and social objectives.

Consider an area within a central business district (CBD). It will have offices, shops, some service industries such as printing, restaurants and entertainment facilities, and perhaps hotels. Valuable though these are they are far from comprehensive. There will be no water harvesting and no waste processing. There will be only very limited effective energy harvesting, and that more by accident than by design. There will be very limited residential accommodation other than hotels and no food production. Probably there will be no manufacturing and very little maintenance. There will be a significant and continuing construction industry with virtually no work-force and with no source of building materials or opportunity for the disposal of building industry wastes. In total it will be far from comprehensive in its array of urban facilities and services. To that extent it will be wholly dependent on the supply of goods and services and people from the surrounding areas. It will also be dependent upon the surrounding areas and the wider environment to accommodate or dissipate or digest its wastes.

Now consider quite a different urban environment and one that typically constitutes a very large proportion of the total area of the built environment. Consider a block of suburban housing – here there is housing and virtually nothing else. There will be the occasional school and corner shop, some recreational space and perhaps a church or community building. As with the CBD area there is unlikely to be any serious water or energy harvesting, waste will be a problem requiring to be disposed of somewhere else. There will probably be no manufacturing and only very limited maintenance (much of it, perhaps, on a self help basis). Food production and some nutrient recycling will be present but only on a very limited scale. As with the central urban area there will be major deficiencies in the local habitat when evaluated as a life support system.

These critiques, focusing as they do on local complementarity of facilities, presume that other aspects of interaction with the environment can be dealt with at the smaller scale, i.e. that individual buildings can be made more energy efficient, environmentally friendly materials can be used and that long-life products that have low levels of environmental impact can be developed. They also lead us to ask ourselves what could be done on both of these important components of the built environment. How could the various urban areas be transformed over time, and quickly enough, to be effective in meeting pressing environmental requirements? Could urban areas be modified and developed in ways that would increase their local complementarity and self-reliance?

The two extreme situations that have been considered would require quite different approaches. Each approach, however, would have to be based on the opportunities that are inherent in each situation. While direct rigorous comparisons may not be practical it can seem that the opportunities for introducing new life support elements within CBDs is likely to be physically limited and further limited by economic considerations. Solar access and energy harvesting are likely to be difficult to provide within existing central urban areas. Any significant amount of food production within already congested areas cannot realistically be envisaged. The space needed for those elements of the life support system that inherently require large areas will not be available. Suburbia, on the other hand, presents quite a different range of deficiencies while its unused or under-used space provides a wide range of opportunities for action.

Some preliminary work has been done taking as a case study a small area (1 ha) embedded in a much wider area of fairly typical Australian suburbia (Roger and Fay, 1991). Built about 1950 as public sector housing at a gross density of 12 houses/ha the area is now partly in private ownership. Originally built to a rather low specification, the density, the general layout and the form of the buildings are not significantly different from what would be found in much of Australian suburbia. While the case study area is near Melbourne something similar could be in or near almost any town in Australia, or indeed in many towns around the world. It may therefore serve as a suitable arena in which to explore possibilities.

The simple questions quickly elicit some simple answers, perhaps the most surprising of which is that the low density suburb is full of opportunity for change. That very characteristic of low density, which seems within the densification scenario to be the source of the environmental problem generating the need for motorized transport, now emerges as the basis of a solution. It is precisely because there is much space between buildings and much of it is under-utilized that a process of infill to increase complementarity in the urban mix can be engaged in, thus reducing the demand for motorized movement of people and goods. It is because there is space available that individual buildings can be easily modified. They can have good access to solar radiation and their surfaces can be converted to solar collectors of various types. The spaces between the existing houses offer opportunity to introduce additional buildings to serve non-residential urban purposes. There are the sites on which to construct new buildings in support of work opportunities, be they production, maintenance or services. Throughout the area there is space for biological production to create food, fish, fibre and fuels. These productive activities can also play a part in dealing creatively and profitably with wastes. All of this can be set within an integrated system of water collection, purification, storage, multiple use and disposal.

There are, of course, restraints that must be applied in bringing a great multitude of urban uses together. It is not without reason that they were previously separated. Frequently, for instance, industrial production can be environmentally polluting by the generation of noise or traffic or the release of noxious effluents or gases. In proposing to bring diverse urban facilities together it is therefore essential that their compatibility is ensured. It can be argued persuasively that for many industries and technologies it is no longer acceptable to release pollution irrespective of location. The environmental imperative of clean production therefore opens the way for reintegration of urban functions.

Throughout this set of propositions on how the suburban environment might be transformed to a sustainable form there has been the underlying assumption that the scale of the various additions could be organized in a way that would be compatible with the scale of existing suburban development. It has been assumed that the various additions could operate effectively at the small scale of the suburban block or individual house site. While it would be ridiculous to assert that everything could be done in this way there are persuasive arguments and a body of economic theory that supports the idea that small is not only possible but also beautiful (Brotchie, 1992; Brotchie, Anderson and McNamara, 1992). This sets out some of the ideas that might be developed within the conventional boundaries of individual sites and in some cases, perhaps, by the consolidation of sites. There are also opportunities in those areas of suburbia which are conventionally held in public ownership. If a reduction in the need for private transport can be envisaged, by bringing demands and supplies more closely together, then the possibility of reducing the space that has to be devoted to roads and associated 'nature strips' may also be realized (Schumacher, 1973). This public space may now be harnessed at least in part in support of the initial objectives of complementarity and local production – particularly biological production.

In all of these changes it is likely that new mechanisms for dealing with the ownership and management of land may be necessary, though much might be achieved by creative use of mechanisms that are already available. In Australia any open spaces between the boundary of a private garden and the roadway, not including the paved pathway, is known as a 'nature strip'. Typically, it is grass with street trees, both of which require maintenance. While they do provide an aesthetic product or ambience they are a net burden on the physical environment.

Actually to effect change to the built environment always requires dedication of resources. We commit capital in support of our aspirations. In the case of the densification scenario it is assumed that the resources come through the financial markets in a conventional way. These market forces focus investment into discreet projects. In the case of the suburban transformation scenario, however, the situation is in some respects rather different. When a process that could, and probably should, take place everywhere throughout the urban area is postulated, not only in one city, not only in one country, but widely throughout the world, it must be accepted that there is nowhere from which to draw the required resources, except, of course, from the local community itself. When the whole urban system is being redeveloped each bit must, on average, be self-financing and must generate its own resources for change.

At first sight this is a daunting prospect until it is realized that this very process of change can provide the opportunity for initiative and creative and productive work as practical responses to specific local social, economic, cultural and environmental situations. The principal resources are the people, their time, skills and capacity to shape their future, plus such conventional resources as they have available to them. In most situations it seems likely that some combination of resources from the formal and informal economy will provide the most appropriate basis for action.

All of this may seem relatively straightforward as an idea but it must be recognized that it is in sharp conflict with many aspects of conventional urban planning. Perhaps the single most powerful instrument of modern planning has been the capacity, enshrined in regulation, to apply zoning. Zoning effectively takes mutually supportive components of the urban life support system and separates them out into isolated urban mono-cultures. In the interests of solving one set of environmental problems, zoning has created the high movement and environmentally degrading urban systems that we now have. The propositions that have been advanced for complexity, multi-use and multiple use therefore challenge directly much of the conventional wisdom. It must be expected that they will meet with resistance from professional, administrative and other vested interests.

The best arguments are illustrations drawn from the case study (Figures 6.1 and 6.2). An individual house is shown in a process of transformation that might be put into effect over a 20 year period. This is combined with a similar scenario for the whole hectare starting

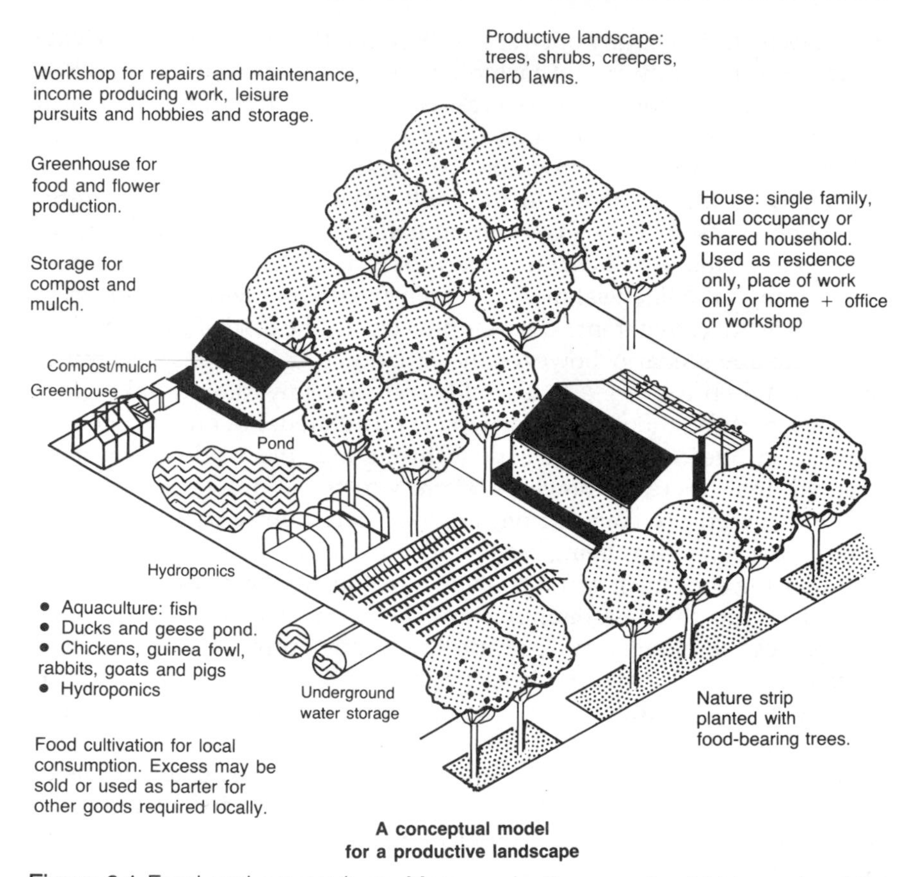

Figure 6.1 Food and aquaculture. More productive use should be made of the sun's energy and rainfall on the site – they are available resources.

originally with the 12 houses. Throughout there is an underlying assumption that the transformation over time can be accomplished by a mixture of conventional financing and self-help. It is envisaged that the benefits of one stage will support the development of later work. While information will be of central importance in pursuing this development strategy the technologies employed would be those that can be appropriated into the life of the community and then employed. Thus, it would be expected to see the development of some traditional skills which would be extended and complemented by the latest research and information technologies. (In Australia, strata title and cluster title systems may be appropriate vehicles. Leasehold systems, the law of tenement and the establishment of covenants to protect the rights and privileges of individuals and the community may all be possibilities in particular circumstances.)

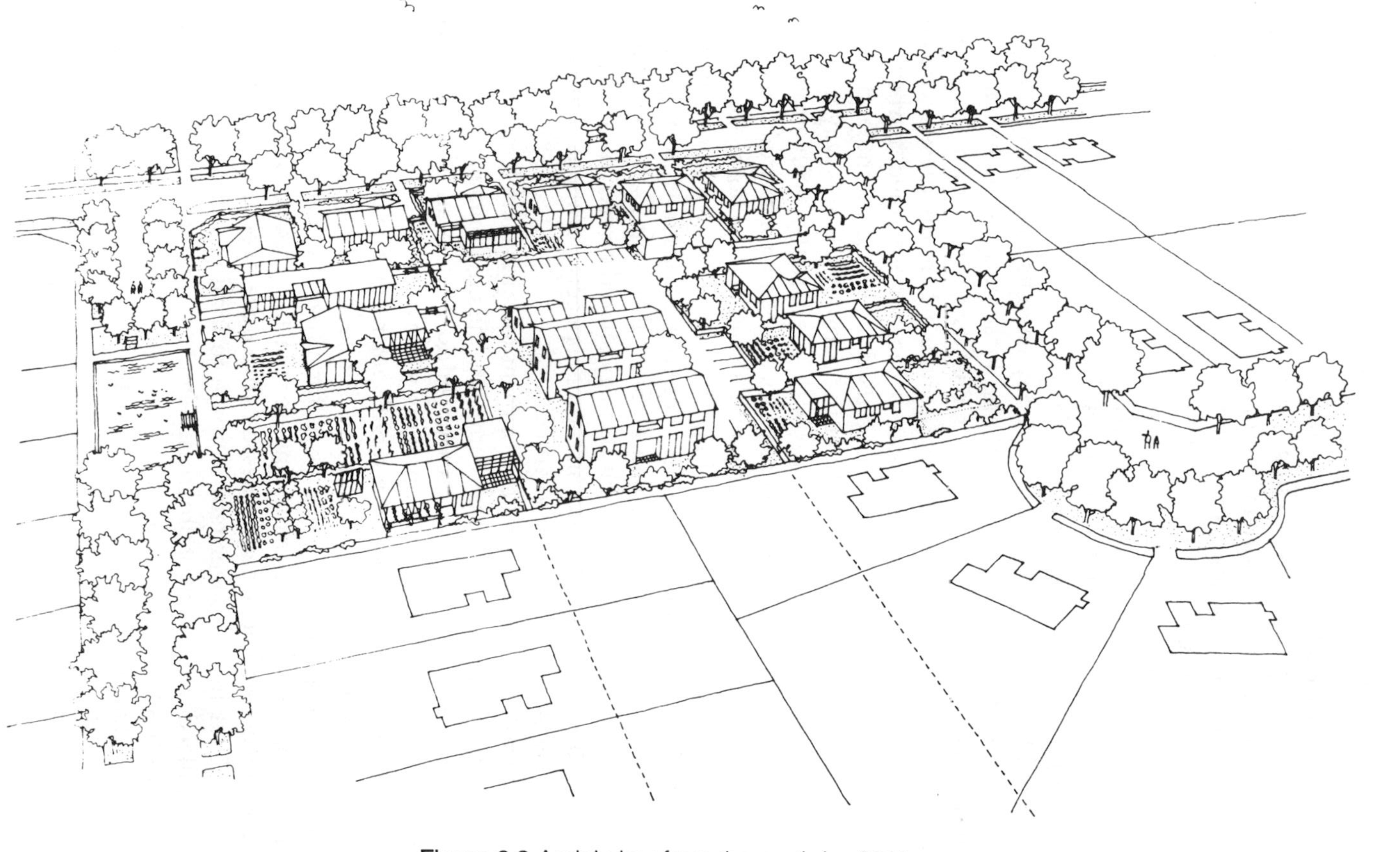

Figure 6.2 Aerial view from the north by 2010.

6.8 Conclusions

The two scenarios that have been put forward are at the extremes of a spectrum of opportunities. In some ways they appear to be in direct conflict and in some important respects they are. Clearly, the ideas of densification and local complementarity can lead to quite different conclusions. To increase the density of housing in a housing area with a view to improving public transport would be exactly the wrong approach if reduction of the need for movement and the development of local comprehensiveness were the objectives. How then can this be resolved? Is one right and the other wrong?

It would appear that each strategy may have its role. Each may have a time scale over which it is relevant and locations and circumstances in which it is particularly appropriate. Consider them in turn. The densification strategy has immediate applications where the present circumstances include under-used efficient urban services, and particularly efficient public transport that could be used to service vacant land either in or near city centres or, less likely, at the periphery. In this way it serves as a strategy that may deal more or less efficiently with growth in population. As such it does little to reduce the environmental impact of those existing urban areas which will dominate the form, operating characteristics and environmental impact of the built environment capital stock far into the future. Its validity therefore is somehow in proportion to the growth of the urban population. Since this is, in itself, an environmentally undesirable situation the strategy should be seen as an expedient capable of being applied in some specific locations until this underlying situation can be resolved.

In contrast, the suburban redevelopment towards complementarity of urban facilities is a strategy that confronts the environmental problems that derive directly from the present stock. This is an approach that produces less spectacular developments yet it does have the potential to mobilize new resources into the development of a sustainable culture and a corresponding habitat system. While the initial gains may be small though many, the potential is for progressive improvement as the environmental opportunities inherent in the present low density of suburban development are harnessed to reduce overall environmental impact. Thus in the longer term a route to local complementarity and the parallel development of progressively more environmentally friendly buildings has the greater potential.

Quite unknown, and unknowable at this stage, is how much change will eventually be required to meet the limiting demands of the biosphere. What can reasonably be concluded is that it would be prudent to retain maximum flexibility by adopting wherever possible approaches that have great potential for the development of convivial and environmentally friendly urban systems. To this end some selective use of densification through medium density infill will serve to deal with some local short- to medium-term circumstances. This should be accompanied, however, by a broader and longer-term strategy of developing local complementarity and local self-reliance.

References

Australian Bureau of Standards (1990) Australian National Accounts, Capital Stocks 1988–89 Canberra, ABS.

Brotchie, J. F. (1992) The changing structure of cities, in *Urban Futures*, Special Issue No. 5, Department of Health and Housing, Canberra, pp. 13–26.

Brotchie, J. M., Anderson, M. and McNamara, C. (1992) Improving the transport system, in *The Proceedings of a Conference on Residential Property Development*, Sydney.

Carnot, S. (1824) *Réflexions sur la puissance matrice du feu: et sur les machines propres à developpée cette puissance*, Bachelier, Paris.

Newman, P. W. G., Kenworthy, J. R. and Lyons, T. J. (1990) *Transport Energy Conservation, Policies for Australian Cities: Strategies for Reducing Automobile Dependence*, Institute of Science and Technology Policy, Murdoch University, Perth.

Robertson, J. (1978) *The Sane Alternative: signposts to a self-fulfilling future*. Published by the author.

Rodger, A. and Fay, R. (1991) Sustainable suburbia. *Exedra*, **3** (1), 4–15.

Rodger, A. and Robertson, G. (1991) Victims, villains and white knights, in *Proceedings of Solar '91* (ed. D. Matthews), The Annual Conference of the Australian and New Zealand Solar Energy Society (ANZSES), Adelaide.

Schumacher, E. F. (1973) *Small is Beautiful: Economics as if People Mattered*, Blond and Briggs, London.

Smith, A. (1961) Of the division of labour, Chap. 1, *The Wealth of Nations* (ed. E. Canan), Methuen, London.

Energy and architectural form

7

Patrick O'Sullivan

7.1 Introduction

It is often believed or perhaps only hoped that there is a simple causal relationship between energy efficiency and architectural shape, and that this relationship, once understood, will enable the designer to design an endless series of energy efficient buildings. Unfortunately, if there is such a relationship, I have not discovered it, and this is because in my view energy is a commodity that building occupants use to enable lives to be made safer, healthier, more comfortable, more efficient, etc. Therefore, the energy used is more associated with what is done in buildings rather than their shape. Of course, some buildings have the ability to use solar heat and daylighting more effectively than others, some need lifts more than others, some more small power, etc. How, therefore, and for what reasons, do people use energy in pursuing their ordinary lives in their ordinary buildings? It is only by understanding this that anything important about this usage can be discovered that might lead to some general rules.

We use energy:

1. To help produce a healthy and comfortable physical thermal, visual, acoustic and air (biological, particulate and gaseous) environment in

Global Warming and the Built Environment. Edited by Robert Samuels and Deo K. Prasad. Published in 1994 by E & F N Spon, London. ISBN 0 419 19210 7.

which to live. (Note: Health and comfort are not synonymous with each other, but of this more later.)
2. To help provide the other necessities of life, namely water, cooking, washing, cleaning, etc., i.e. the 'power' for labour saving devices.

The quantity of energy needed to accomplish these tasks is predominantly culturally dominated, for example compare the domestic energy usage of North Americans and Northern Europeans, although with a strong climatic bias.

Perhaps more importantly the vast majority of people live in buildings, and in Northern Europe spend over 90% of their time in them, so that it is not surprising that design theories have been developed which help to relate energy usage to the way in which lives are spent, of which the two most important are climatic modification and the 'lack of discomfort' theory.

In this chapter I will try to explain how these theories allow one to design the fabric of the building to modify the external/internal climate in an energy efficient way to produce a healthy, comfortable and cost effective environment. Furthermore, the general rules that emerge from the application of such theories to a number of designs – rules that emerge from arguing from case study to causality, will be discussed.

7.2 The theories

As mentioned above, the energy needed to support us in our lives in our buildings is dependent on two factors: our culture which influences the conditions under which we are prepared to live; and the climate, which determines the importance of buildings in our lives.

7.2.1 Our culture

In the culture of which I have experience, i.e. the Northern European, over 90% of our time is spent in buildings, and therefore high standards of health and comfort are demanded of them. Basically, the higher the standards and the longer the usage, the greater the energy consumption. In Europe the energy usage per capita is high. Approximately 50% of delivered energy, i.e. the energy delivered over the boundary of the product or process, is used in buildings, and 50% of that (i.e. 25% of the total) in our homes. There is therefore a great incentive to make our buildings energy efficient.

The design theory used to relate people and their functions to buildings and their energy usage is known as the lack of discomfort theory. In this theory the results of many decades of comfort research are used to define (in terms of the physical environmental parameters) the zone within

Figure 7.1 The parameters leading to personal comfort or discomfort – the lack of discomfort theory.

which the body is least responsive to changes in the physical environment. In the UK, for example, this approach leads to ranges of air temperature of 19–23°C, of relative humidity 30–70%, of 'light level' 400–1000 lux, of air movement 0.1–0.5 m/sec, etc. It is useful to plot these parameters on a single diagram, see Figure 7.1. This shows that there are two zones related to personal comfort. A clear zone of discomfort (too hot, too cold, too dark, etc.), which is to be avoided, and a lack of discomfort zone where the body is least responsive to changes in the physical environment. It is the creation of this latter zone which needs to be achieved in the interiors of buildings.

In the UK less energy is used if the room air temperature is kept at 19°C rather than 21°C in winter, the lighting level at 400 lux rather than at 1000 lux, etc. In other words, there is a simple relationship between the standard of the physical environment and the energy usage needed to sustain it. In fact this lack of discomfort theory is the basis of most European countries' building energy legislation and savings programmes.

However, the lack of discomfort theory is even more fundamental, in that if there is one physical parameter that appears to act as a primary correlate with personal comfort/discomfort, it appears to be air movement and its associated air quality. If an internal space is small, for example a domestic living room, merely opening the windows and internal doors will normally ensure that the air flow through the room is sufficient to maintain the lack of discomfort zone. As the internal space becomes larger, there comes a time when the size of the space is such that merely opening the windows and doors no longer suffices. The reason being that the air no longer moves across the room, but rather crawls around the walls leaving a zone of air movement near the window, but none in the middle of the space. To maintain the lack of discomfort zone in such a space, the air now needs to be mechanically driven through it. In other words there is a relationship, or more properly a set of relationships, between the size and complexity of the spatial arrangements of a building, and the order of technology, and therefore the order of difficulty, and therefore the order of energy usage, and therefore of cost, necessary to maintain the lack of discomfort zone in the building.

This then is the theory at its simplest. In reality, of course, buildings have to be healthy as well as comfortable, and unfortunately (as with diet) comfort is not necessarily the same thing as health. As a result, the simple lack of discomfort theory has to be increasingly modified to take the overridingly important health factors into account. For example, experience shows that as the air temperature (from 19 to 22/23°C) and relative humidity (from 45 to 65/70%) in a building rises, so too do the health risks. Lower air temperatures (19/20°C) and more clothes are healthier in winter, but not necessarily perceived to be so comfortable. The very 'power' of this comfort based theory has led us in the past to design simplistically 'comfortable' environments, within apparently energy efficient enclosures, which, in some cases, perhaps as a result of changing standards, have not always proved to be healthy. For example, simple energy efficient 'natural' ventilation systems, based on the concept that unlimited quantities of 'fresh' external air are available (to dilute the external and internal airborne pollutants to safe levels) can no longer be expected to produce satisfactory results in highly polluted high density urban areas.

Charts similar to that shown in Figure 7.1 can be drawn for any culture, and for the range of air and water based contaminants that exist in particular locations within any cultural grouping. However the principle always remains the same, namely: firstly, determine the internal health, and thus the 'comfort' standards, that are required, and not the other way around; secondly, determine the spatial planning, i.e. the spatial size and complexity of the building within which the standards are to be supported. This will provide the clues necessary to determine the order of complexity of the technology necessary to maintain such standards, and therefore the energy usage. Simple spatial relationships, where the individual spaces are small, where the outside air in individual spaces is 'clean', and the external and internal environments thus quiet, light and airy, will normally be the most energy efficient and healthy. However one must accept that the results so produced will not always be the most acceptable, or indeed possible. One of the great unresolved issues is which group of people has the rights over the limited dilution properties of the external air. Is it for example for the traffic engineers to use up, or must some be left for the building environmental designers? In which case, how much? Until this and other such critically important issues can be resolved, one can only act with sense and caution.

7.2.2 Our climate

The theory of climate modification argues that we can measure, to some extent predict, modify, but unfortunately not control the climate external to our buildings. The theory further argues that people work well in buildings in reasonably well understood healthy and comfortable environments, this latter as a result of the correct application of lack of discomfort theory. It next argues that a, if not the, prime function of a building is to modify the existing external climate to the preferred internal one, and that the building can help in this process in three ways:

1. The mere fact of constructing a building on a site alters the climate around it, either for the worse or for the better. The effects of buildings on site climate can be quantified and used to make the building more energy efficient. In Northern Europe this process of microclimatic design can be and has been for many years simplified into one major and important rule. This is to design the urban spaces to allow the sun (solar radiation) into the spaces surrounding the buildings, while at the same time to keep the wind out (Figure 7.2).

   'Success' in external climatic design in Northern Europe is measured in terms of the increase in both the level and stability of the external air temperature that results from this design and construction process. An increase of between 2 and 4°C can usually be achieved in temperate climates, e.g. see Figure 7.3. However, much higher temperature differences, of the order of 6/8°C, have been recorded between the centres and outskirts of very large cities such as London. The increase in external air temperature reduces the conductive heat loss from the building in winter, effectively shortening the winter heating season, saving energy incidentally. Correct microclimatic design can also considerably reduce the convective heat losses by reducing the chill factor,

Figure 7.2 Radiation and wind force balance.

also saving energy. Different but similarly principled techniques can be applied to microclimatic design in other climatic conditions.

2. It is the actual construction of the fabric of the building that acts as the next stage of the climatic filter/modifier. Buildings are in effect made up of 'sandwiches' of materials of differing properties combined together in the construction. The combination of materials determines the quantity of heat, light, sound and air that enters and leaves the building, at what rate and over what time. Therefore, the design and construction of the fabric plays a large part in the energy efficiency of the building.

   There are two approaches to the question of fabric design, namely:

   (a) Climatically rejecting fabric design. In this case, as the name suggests, the combinations of materials that make up the fabric of the building are designed to reject the external climate. Climatically rejecting buildings are the buildings of cold climates. They are, and certainly should be, designed to be light, airy (large volume per person), thermally efficient and thermally stable. The roofs are pitched with overlapping slates or tiles, to reject the rain. They are ventilated through the eaves to remove the damp, an often difficult constructional detail to design correctly, and more recently insulated with fibrous materials, usually just above the ceiling

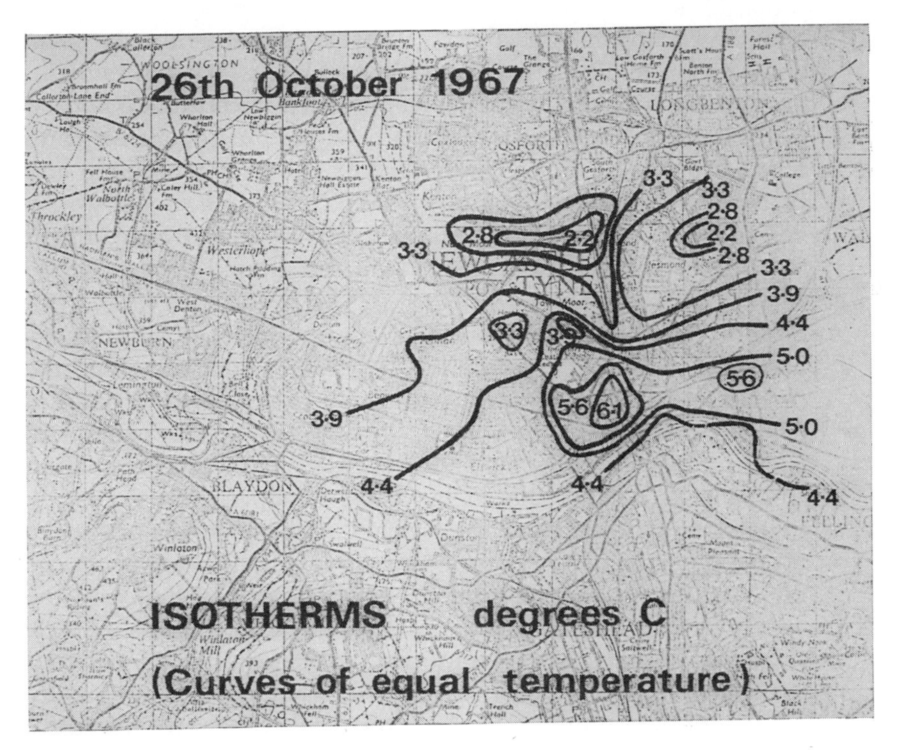

Figure 7.3 Curves of equal temperature.

level, to keep the heat in and the cold out. The walls are of massive (stone, brick, wood) construction, with an outer leaf to reject the rain, a ventilated cavity to prevent damp penetration, an insulating layer again of fibrous material and an inner leaf. The outer leaf absorbs and releases the incident solar radiation, thus improving the quality of the external environment. The insulation prevents the inward and/or outward passage of heat. The inner leaf, which absorbs and releases the internal heat, smooths out the internal temperature swings. The inner lining of the inner leaf, for example the type of wallpaper, can be used to improve where necessary the internal radiant balance, so important if internal temperatures are to be kept low. Designed correctly, such a thermal constructional sandwich can be very effective and energy efficient. The windows, which typically occupy up to 40% of the external walls, are single glazed and are located, shaped and sized, to optimize the view and the penetration of daylight. The advantages of well designed, glare free, daylit spaces that do not suffer from frequent solar overheating during the hours of occupation are impossible to overestimate. All consumer surveys ask for light, simple, airy spaces, that the occupants can control. They are consistently the most energy efficient. The air which enters and exits through the windows is traditionally large in volume, often uncontrolled and no designed ventilation exists. This is partly to provide air for combustion, partly to dilute contaminators and reduce odours, and partly because it was thought (often correctly) to be healthy. Any designed system was felt to be unnecessary, expensive and difficult to achieve. The floor in contact with the ground is often uninsulated.

Since 1974 the need for increased energy efficiency in buildings has developed, and the building legislation and, therefore, the process of building design construction in Northern Europe has developed as follows. Initially it was, and to a certain extent still is, one of increasing the insulation standard of the major building components, namely the roofs, walls, windows and floors, by mutational changes to the 'sandwiches' of the materials that make up the construction. The purpose of these actions being to reduce the heat loss through the fabric as a whole, and therefore the energy needed to maintain the internal environment at a suitable standard. This process was adopted so that the traditional constructional types and processes could be maintained. It was rightly felt that changes to the basic constructional types would, if introduced too quickly, result in many constructional failures, and, what is more, failures associated with damp. At first therefore the job was to decide how existing constructional sandwiches could be improved. Use as much insulation as the existing construction can take, became the motto. Initially a combination of the versions of old products emerged, e.g. 'insulated' blocks. Later new materials emerged, e.g. the whole range of 'soft' insulating products such as mineral fibres and boards, double, triple and gas filled glazing, etc. Finally a range of new constructional techniques such as filled cavities and highly insulated timber frames came on the market. The situation is now that the problems of a highly insulated fabric for most if not all situations have been solved and are no longer an issue. Standard solutions exist in every country and should be adopted by every designer.

As the fabrics became more highly insulated, the main energy usage became associated with ventilation, and its attendant heat losses and gains. For the first time the careful design of the ventilation system in a building was recognized for what it is, namely a, if not the most, critical component of healthy energy efficient design. Buildings, as well as leaking energy through inefficient ventilation design, also have the unfortunate ability of both producing and concentrating pollutants in their spatial complexes. Over-ventilated buildings can be very unhealthy!

The era of energy efficient and healthy designed ventilation had arrived. In a house in Northern Europe for example, this would consist of: the air entering each occupied space in winter via trickle ventilators set in each window in each room, moving through the house via cross flow ventilators acting as transfer grills in all the internal doors or door frames and 'natural' or mechanical extract from the bathroom and/or kitchen. The specifics depending on the particulars of the dwelling design. In summer the windows would be opened thereby reducing the static pressure head on the system and increasing the air flow. Such designed ventilation through volumetrically correct spaces would not only limit energy usage, but also reduce pollutant concentrations, mitigating against the effects of damp and mould growth. However, it is important to remember that the understanding of the effect of airborne pollution on ventilation design in high density urban areas, in energy efficient housing, is still only in its infancy. The overall assumption is still that if the outside air is not suitable and sufficient for pollution dilution then the health regulations will prevent the construction of houses in that location. In more complex non-domestic situations the same rules apply, although the solutions are also more complex. Nothing puts up the energy cost of the ventilation system, and increases the health problems, more than reducing the volume of air per person, and needlessly increasing the spatial complexity. A good innovative engineer can ventilate almost anything, but at an energy and health cost.

In non-domestic buildings, once the fabric and ventilation designs have been resolved the energy cost is normally dominated by the lighting design. Suffice it to say that over recent years windows have varied in size from 20 to 40% of the external façade as opinions on the relative importance of saving heat and optimizing daylight have waxed and waned, and single glazing has given way to double or triple, air or gas filled, again depending on the perceived view of the relative importance of heat and light. However, as the production of CO_2 has joined the energy question in our minds, the advantages of daylighting are difficult to refute.

It has been simple, but not easy, to determine which were and are the correct design actions to take in terms of saving energy by fabric design. What has not been so clear is how far it has been worth going, as this can only be determined by the value society places on its energy supplies, and on the environmental pollution associated with its energy usage. There are basically three ways in which a society can determine such a value:

(i) By manipulating the various parameters in terms of a view of cost-effectiveness – cost of measure vs. energy saving – of, for example, roof insulation, ventilation, heat loss, etc. The difficulty of this system, which is currently the most commonly used in the construction industry, is that if energy prices rise year by year and/or the environmental cost of using the energy increases one is never achieving enough, and vice versa.

(ii) In terms of some standard of acceptable energy usage of, for example, energy per square metre, for each building type. The difficulty in this system which is commonly used by building owners and their facility managers, is that a too optimistic approach could quickly bankrupt sectors of the construction industry, and a too pessimistic view would mean that little was achieved.

(iii) By the application of particular views on interfuel substitution to building energy usage via the use of useful, delivered or prime energy. This methodology applies particularly when the availability of energy supplies to a country and/or building are limited or skewed, e.g. 'gas is good for heating and electricity is bad' etc.

In reality most European countries use a mixture of all three, as modified by the current political view on the strategic value of indigenous energy resources, plus a risk analysis of the attendant issues, e.g. health. Sanity is preserved by ensuring that no building or energy regulations act retrospectively. Suffice to say that most fabric regulations have been in place for some years now, and the focus has moved to the more difficult area of ventilation and lighting control with all the attendant health hazards.

Climatically rejective design is still the most common, simplest and most straightforward form of energy design. It is felt to be applicable to all building types and has the advantage that once the building has been constructed it is thermally stable and energy efficient. It is therefore felt to be robust against climatic change and it minimizes the effect of energy price rises and pollution regulatory changes. Such a building will, however, probably use more energy per unit area than a properly designed climatically interactive building, but a lot less energy than a badly designed one. It therefore represents the fail safe option.

(b) Climatically interactive design. In this method of design the external climate is not rejected from, but rather 'invited' into, the building in the expectation that by this means the energy usage of the building will be reduced. The reduction, it is hoped, will occur by displacing some of the fossil fuel produced lighting and heating with 'solar displaced lighting' and 'solar displaced heating'. It depends for its success on a clear understanding of the climate concerned and a clear view of any future climatic changes. If, for example, the solar radiation falling on the UK were to increase over the years as a result of holes in the ozone layer, global warming, etc., then perhaps climatic rejection techniques would be the safer and the healthier option.

The basic rules of thermal efficiency and stability that applied to climatically rejecting building are still relevant, but now, instead of distinguishing between two climatic zones, i.e. inside and outside, three zones are distinguished. These are the internal, intermediate and external zones. The internal zone is the core of the house, namely that part to which one 'retreats' into during the winter – this area is actively conditioned in that it is fossil fuel heated and to a certain extent fossil fuel lit. It is designed on traditional climatically rejecting lines. In spring and autumn the intermediate space – the sun space – the conservatory, comes into use. The smallest and probably the most effective sun space is the bay window. This space, which can be shut off from the core space in winter, is heated and lit by the sun only and provides extra living space for up to nine months of the year at no energy cost. In this way the energy using winter space is kept to the minimum. In the summer the external space – the patio/garden – comes into use, again at no energy cost. Clearly, if the intermediate space is ever heated with or lit by fossil fuel derived energy then the energy usage increases dramatically and the design system fails.

In Northern European domestic premises the intermediate space is normally designed as an unheated conservatory or sun space, oriented south of an east/west line capable of being effectively thermally isolated by, for example, doors from the remainder of the house. In educational and commercial buildings the sun space may take many forms of which the most common is the unheated atrium or linking space. As such, in the winter it is used for communication and storage only. Attempts to use such spaces during the whole year, with the attendant heating, cooling, ventilation and lighting that this implies, can incur enormous energy penalties, and as such must be treated with extreme caution, no matter how attractive such spatial usages initially appear.

In this type of building the energy bill for lighting usually dominates, assuming sensible fabric and ventilation rules are understood and followed. To reduce the energy for lighting the principle is to spatially plan the usable spaces so that as large a number as possible are associated with the intermediate zone.

Then, design the spaces on the environmental principle of letting the daylight in and keeping the summer sunlight out. The so called LT method of design.

Climatically interactive designs are still relatively new in Northern Europe, and will always be more difficult and carry a higher risk of failure. For example, if an unheated conservatory is subsequently heated by fossil fuel, more energy will be wasted than saved. Also the method is not applicable where space is itself at a premium and must be used twelve months of the year.

3. The final stage of the climatic filter, which is not the subject of this chapter, is the fine tuning of the internal climate produced by the interactions of the microclimatic and fabric designs. This is the design of the internal heating and lighting systems, the interior design that supports the lighting design, the radiative balance of the surfaces, the arrangement of the furniture, etc. In this area, whilst there are codes of good practice for lighting and heating design, regulations are only now beginning to emerge.

7.3 Conclusions

In this chapter I have tried to show that there is no simple relationship between energy and architectural form; the relationship is rather with climate and culture. Furthermore, as energy considerations are, at best, only one of the factors to be considered in the improvement of humanity's lot, the best solutions are always the simplest.

Environmentally benign architecture: beyond passive

8

Jeffrey Cook

8.1 Introduction

Everywhere in the industrialized world the same issues have emerged concerning the built environment. But in each country, in each region, or even each city, the mixture and emphasis of concerns varies. Thus beyond the anonymity of the international style in the industrial world is the particularity of personal and professional responses in the post-industrial era. Each of us, from our own particular geographic and political location, must develop particular insights about the dramatic adjustments necessary to practice responsibility now and especially in the future.

Reviewing the decade of the 1980s, it is evident that architecture as practised in the USA is in retreat. Particularly in the recognition of responsibilities in energy and environment, and in the application of both current knowledge and available technologies, architecture is currently out of touch. More than ever, appearances pre-empt performance; expedience and low first cost have overshadowed reasonable life cost; and buildings continue to be designed to the values of the expansive 1960s. The attitudes of the 1980s were thrown into sharper relief in 1991 by the willingness of the USA to go to war for oil halfway round the world, as

Global Warming and the Built Environment. Edited by Robert Samuels and Deo K. Prasad. Published in 1994 by E & FN Spon, London. ISBN 0 419 19210 7.

well as by a proposed national energy plan that would accelerate the capture of our already sparse petroleum reserves.

In contrast, is the simultaneously expressed public concern in the USA with the quality of the environment. In repeated polls it is among the current major shifts in public awareness and concern. Environment now occupies a high priority as citizens realize that life without clean air and water is not only unpleasant but untenable. A national public opinion poll conducted in December 1990 in the USA by the Union of Concerned Scientists and the Alliance to Save Energy found that 47% of the public would give top priority to government funding for renewable energy research and development (R&D), 28% would give top priority to energy efficiency, 12% to oil and coal, and just 10% to nuclear energy. Those trends continue in the face of economic recession and uncertainty.

In a letter to (then) President Bush on 7 December 1990, the Union of Concerned Scientists and 17 other environmental and energy efficiency groups pointed out that 'numerous studies and considerable testimony presented to the Department of Energy show that greater energy efficiency and renewable energy sources comprise our largest, cleanest, lowest cost, near-term energy resource'. The letter called upon the Administration to set a goal of improving US energy efficiency (energy/GNP ratio) by at least 3% a year, and of obtaining 20% of its energy from renewable sources by the year 2000 (up from the current 8%).

Yet even with a new president the US federal administration has not changed its retrogressive policies. It was 'business as usual'. The universal inertia of politicians to energy and environment consciousness is perhaps best represented in the USA. Certainly the dramatic suddenness of political shifts in Eastern Europe and the USSR demonstrated again how politicians convince themselves that holding the system firm in the old ways is not only blind to emerging realities, it also accelerates and exaggerates the system adjustments that inevitably must be made. In energy environment policy, the necessity and urgency for change continues to be documented in the USA by Earthwatch and many other autonomous agencies.

8.2 Passive solar architecture in the USA

In the USA awareness of the potential of the sun to energize buildings had reawakened well before the Arab oil embargo of the winter of 1973–74. The professional responsibility of common solar sense in that generation was reborn with a new label, passive solar. It was not a revival of the old solar movement of the 1940s and 1950s where the sun was increasingly seen as a substitute fuel with its own increasingly complex hardware. Rather it was a genuine fresh start concerned with renewables that generated its own heroes, new architectural prototypes, and new literature.

The invention called 'passive' was a reconceptualization of architecture as the climatic modifier. It went beyond getting the windows in the right place both for direct solar heating and for cross-ventilation. It proposed the whole building and all of its parts as contributors to the thermal balance of comfort. It described space conditioning driven by natural phenomena interacting with architectural elements. Its avoidance of motorized devices to move fluids and control the interior conditions minimized resource conversion both on and off site.

By the late 1970s passive concepts appropriate to single family houses had stimulated a reconceptualization of larger buildings with an appropriately different matrix of passive strategies for office buildings and other building types. A series of state office buildings in California completed in the early 1980s best exemplified that there is no predictable passive or solar style. Passive strategies allow a variety of aesthetic opportunities. Yet from that impressive position at the beginning of the 1980s in the USA we have slipped back in the recent decade to a general mentality of the mid-1960s when fossil fuels were so cheap and available that solar applications disappeared.

8.3 Missed models

The most promising historical example of an integrated passive heating and cooling system, the Skytherm System of Harold Hay 1967–68 (Figure 8.1), and the most stunning architectural image of the New Age, Steve Baer's Zomehouse 1971 (Figure 8.2), were both well in advance of the new trend of the 1970s. Their performance records in rigorous climates providing between 70% and 100% of the space conditioning energy for both heating and cooling set reasonable standards. Today in most climates and with many systems over 100% performance is designed for, to provide a cushion for the variability of climate.

Both Hay's and Baer's demonstrations were simultaneously naïve and sophisticated at many levels and yet were eminently understandable. Both were exclusively the products of highly independent and totally unsubsidized inventors. Neither was an architect or an engineer in the sense of being professionally educated or registered. Both had presented aesthetic integrities and thermal sensitivities in advance of the first Earth Day of 1970. Both seemed all the more vital on the 20th anniversary of Earth Day in 1990, although both had been only intermittently influential in the intervening two decades. Neither building still exists in its original form.

Yet by the 1972 United National Conference on the Human Environment in Stockholm the passive solar movement in the USA was well

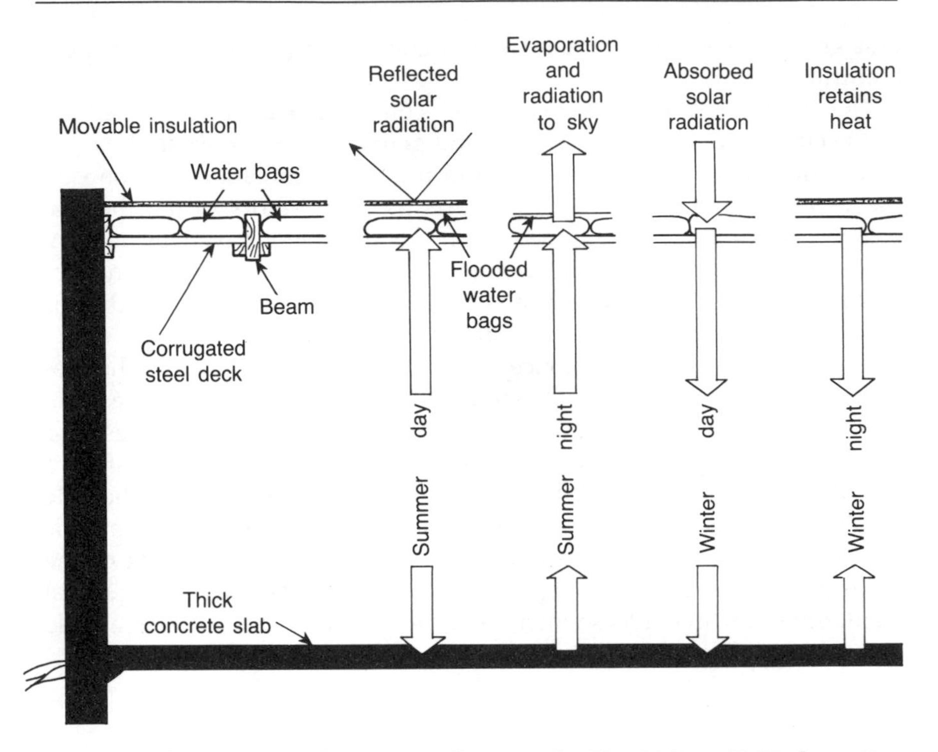

Figure 8.1 'Skytherm' or 'Thermo-pond' system by Harold Hay, 1967. Operating principles for integrated 100% passive heating and cooling.

underway totally independently. Inspired by Hay and Baer and many other influences in the challenging climates of northern New Mexico, that passive movement has ultimately informed responsible architectural practice globally.

8.4 Architecture of the 1980s

During the 'me' decade of the 1980s, the oil glut and resulting bargain prices for petroleum reinforced the evaporation of social responsibility in the USA. At the same time the 'sick building syndrome' emerged and indoor pollution began to match outdoor pollution in the most advanced nations. In the USA the Environmental Protection Agency estimates the cost of indoor air pollution to be as much as US$1 billion a year in medical expenses, absenteeism and general damage. Thus while the built environment became more stable structurally, it became less stable and more threatening environmentally. By threatening all living things it visibly began to endanger humanity itself.

Simultaneously solar and bioclimatic research and professional devel-

Figure 8.2 Zomehouse, Corrales, New Mexico by Steve Baer, 1971. (Photo: Cook.)

opment investments dramatically declined in the USA, but accelerated in Europe. Anticipating the common need for imported oil and the growing environmental pollution that would also have to be shared, the European Community (EC) funded activities such as handbooks, competitions and calculation methods as well as continued basic research for both architects and building scientists. Unlike the jerky passive solar initiatives in the USA in the late 1970s the EC programmes are multinational, multi-staged, multi-year and interdisciplinary. Their depth and continuity assure long-term results both in terms of common performance standards in designing buildings as well as new educational tools and standards in schools for the new generations of professionals.

By the early 1980s in the USA leading architectural historians and theorists had published revisionist histories of recent architecture. The revisions included broadening the geographic range in the late-20th century beyond the USA, Europe and Japan to include many parts of the rest of the world – parts primarily in overheated climates, as well as in developing locations where resources are less plentiful. At least as critical for historians and for writers of theory was this broadening of design criteria to include issues of climate, comfort and appropriate materials for construction. However, the impact of these revised academic standards in professional schools of architecture may take a generation to reach currency in architectural practice.

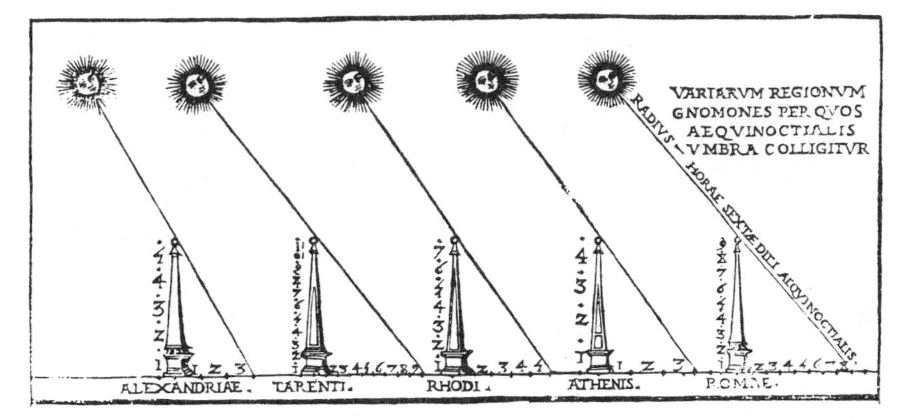

Figure 8.3 Early science in architecture: sun angles for five Mediterranean cities, 1521 (Como Vitruvius).

8.5 Architecture as science

Modern architecture in the 20th century emerged as an industrial artifact and as an aspect of universal commercialism. Simultaneous was the initiation of a new scientification within the design process. Architecture has always been part science, but that part has been growing in the Western world. Since the invention of the printing press and the Renaissance focus on bookish understanding, common knowledge began to be abstracted and set in print. For instance, an early Renaissance illustration that shows the different sun angles for five Mediterranean cities is from the most beautifully illustrated edition of Vitruvius, the Como edition, 1521 (Figure 8.3).

Architecture has also always been part art. Thus the modern reduction or elimination of accepted iconic elements and of traditional subjective responses has encouraged a freshly objective examination of the origins of architectural forms and fragments. By the 20th century certain traditional rules such as the slope of roofs or the length of overhangs were subject to scientific examination and reduction. Rules or even building codes began to formalize what once were intuitively derived and common practice conventions. Such physical determinism was crude in its ignorance of many human and cultural issues even if it was intellectually sophisticated. It sometimes passed for science, since it was not art.

Threads of the origins of 'bioclimatic design', perhaps a more comprehensive term than passive cooling and heating, can hardly be identified here. Our architectural historians have not isolated those distinct concerns that collectively had established the scientific basis for a climatically responsive architecture before the Second World War. In effect they were a codification of common knowledge and traditional wisdom based on the accretion of learning by doing and living with the results. It is obvious

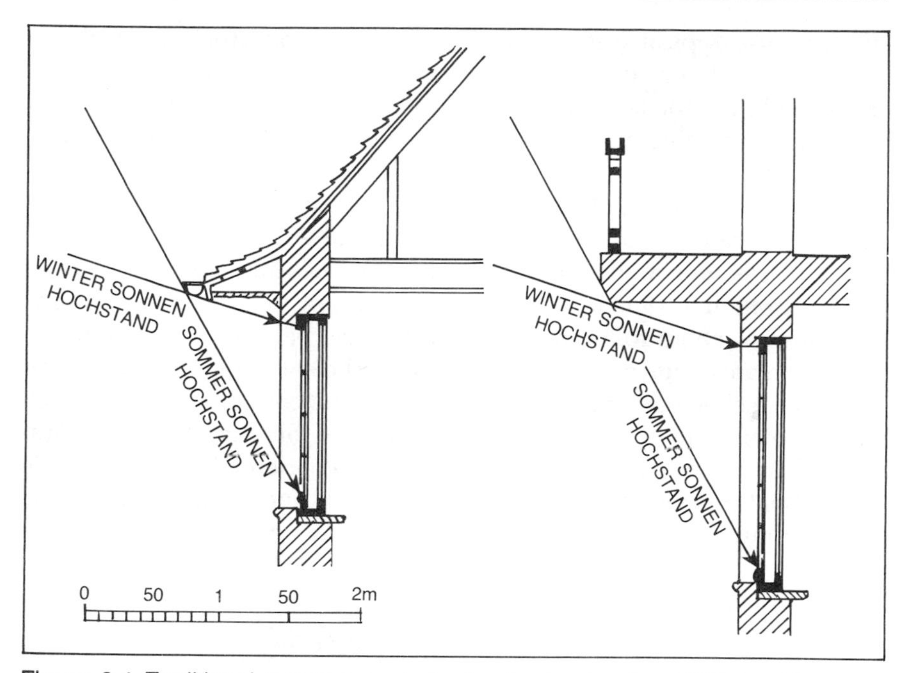

Figure 8.4 Traditional passive control of solar heat gain, 1925, by Hermann Muthesius.

that no individual and no institution was the prime mover. Many people in several parts of the most industrialized countries initiated a diffused effort that continued internationally to create formalized bioclimatic information in a diffuse manner.

For instance, in an otherwise conventional book on the design of the single family house published in Munich in 1925 is a diagram (Figure 8.4) that clearly gives overhang dimensions for solar control over windows. Does this represent a new invention? Or is this simply the publication of a traditional rule that previously was knowledge of the trade and thus had been transferred orally?

8.6 Architecture as package

Architecture and life in the industrial age have become segregated and regularized into predictability that can be bought and sold. The urban experience has been sterilized by monocultural zoning. Architecture has been reduced to sets of building topologies. Reductionist architecture is composed of subassemblies of roof, curtain wall, structural floor, drop ceiling, etc., abstracted fragments of sheets and sticks all industrially produced to uniform standards. Among the most refined examples, 860 Lake Shore Drive Apartments, Chicago, Illinois, by Mies Van der Rohe,

1951, is also typical (see Figure 8.5). In this architecture, comfort and environmental control get assigned as a requirement of one of these subassemblies, not as a responsibility of the entire system. Passive solar heating and passive cooling, especially, have suffered from this reductionist mentality.

The mobile market-place, by providing the cheapest price, has also cheapened value. Thus aesthetically by the late–20th century, all human-made places look alike. Industrial culture is presently characterized by the international trivialization of human life. For instance, the glass box continues to be reproduced as an architectural package design solution in all climates and in all locations as a symbol of being modern, as an image of technology and as a celebration of the control of power which is irresponsible in its wealth. An examination of this current architectural artifact from any scientific, intellectual or experiential criteria reveals how trivial this response is in comparison with known alternatives. The growing dependence on mechanics and plumbers proves how fragile the

Figure 8.5 Reductionist architecture at its most refined: 860 Lake Shore Drive, Chicago, Illinois, by Mies Van der Rohe, 1951.

maintenance of this precious and indulgent performance. The growing awareness of global environmental degradation reveals the irresponsibility of such an ethic. But if all glass buildings could be designed to magically respond thermally within the designed arrangement of their molecules, so heat would be admitted or rejected as required, would such an all glass aesthetic express the sensitivity of the built environment to the natural world that must become standard in a sustainable future?

8.7 Evolving environment standards

The industrial era began with the gross deterioration of standards for human accommodation. In parallel with the degraded workplace, housing in the early industrial period was polluted, disease-ridden, inhumane and often physically destructive. These conditions still persist as a product of industrialization, but now they range globally, and are especially prevalent and too often excused at early stages of industrialization.

Earlier standards have been compensated for through the gradual provision of totally controlled interior human environments, at least in the most industrialized countries. Increasingly the destructive conditions have been moved outdoors. The result has been to spoil the immediate adjacent natural environment by creating sealed interior envelopes of mechanical environmental control. It is also a design strategy of gross and aggressive dependence on major resources that are remote, and thus out of sight. Destruction through mining or electric power plants is more easily ignored when it is in another county.

The totally controlled interior environment has been achieved by the crude conversion of fossil and nuclear fuels to provide artificial light, artificial heat, artificial cooling and artificial ventilation. The lowest denominator architectural elements of industrial culture are the light bulb or fluorescent tube, and the air-conditioner or vapour compression heat pump. The architectural enclosure to support these elements is the smooth sealed building and its parasitic fuel driven mechanical equipment plant. Occupants of these buildings have little control inside their industrially packaged environment, except to instruct the thermostat or bank of lights that may control a whole floor, regardless of size.

The cultural change associated with the post-industrial era or with a more energy efficient and pollution free economy should not be the elimination of artificial lighting and thermal conditioning, but the miniaturization of these artificial sources to allow individual and local control and to reduce dependency. Simultaneously there is an increased appreciation of the need for more direct human connection to the natural environment. Daylighting is a natural extension of passive solar heating and cooling. In larger buildings, daylighting is considered the prime cooling strategy by displacing the heat and energy use of artificial lighting. Daylighting is increasingly a major attribute of important public buildings of international standard regardless of energy or environmental concerns, a reinforcing trend that simultaneously reconnects us to nature through architecture. But when architects are informed that seven bricks require an energy expenditure equivalent to a gallon of gasoline, will they continue to think of the fabric of a building in the old way?

8.8 The indoor air industry

The market for products and services to prevent or correct indoor air pollution is suddenly big business. According to a 1991 report, US Non-Industrial Indoor Environments, 'the total market for indoor air quality (IAQ) consulting and analytic services, radon detection products and indoor environmental monitors, which was estimated to be $248.4 million in 1989, will be 10 times this size by 1996 – a whopping $2.5 billion'. The report also says that when the money that will be spent for new building control systems and to retrofit heating, ventilation and air-conditioning (HVAC) systems is included the IAQ figure for 1996 will total almost five billion dollars.

'Much of the development of non-industrial IAQ problems,' says the report, 'has arisen since the energy crises of the early-1970s, when buildings were weatherized and made more energy-efficient.' The result has been the so-called sick building syndrome, in which a building's occupants suffer symptoms such as headaches, fatigue or eye, nose and throat irritation. Indoor air pollution can even be deadly, a point made dramatically when the bacterium of Legionnaire's disease from an air-conditioned Philadelphia hotel killed 34 people in 1976.

Increased public awareness of IAQ issues has led to a boom for the developers, manufacturers and distributors of IAQ products and services, and for investors. Revenues for new building control systems and retrofit HVAC systems in the USA were an estimated $1.17 billion in 1989, and were projected to grow to $1.32 billion by 1990, $1.96 billion in 1993 and $2.49 billion by 1996.

At the same time, the report warns, new legislation on indoor air quality is proliferating, along with new lawsuits. A 1990 lawsuit in Missouri awarded $16 million to a family that claimed to have suffered 'chronic formaldehyde poisoning' from the materials used to construct their home. Lawsuits involving radon have prompted real estate companies to routinely check for radon levels before placing homes on the market.

The report predicts that new legislation in the USA in the 1990s will probably mandate indoor air quality standards, just as outdoor environmental standards were mandated in the 1970s and 1980s. It predicts that today's heightened environmental sensitivity and accelerated mass communications will bring this new legislation to pass more quickly than the earlier legislation and that most of it will be in place by the mid-1990s.

8.9 The environmental industry

The indoor air industry is symptomatic of a mind set that continues to cripple all advances in building technology. The inability to understand whole systems and the enthusiasm to treat only one part of a building continues to breed new specialists who then become even more myopic because of their focus. The traditional role of the architect as the master designer who integrates all the technologies becomes more difficult as technologies and specializations proliferate.

The environmental industry in the USA as represented by national conferences and trade shows is an accelerating and thriving business. It is also a large scale hardware-heavy band-aid effort. Four of the five goals of a typical major national event in the USA such as an environmental conference and exhibit are 'remediation', 'treatment', 'compliance' and 'recycling'. Only the fifth, 'prevention', might be identified as a shift of goals, and even here the attitude of the environmental industry is one of avoidance rather than reconceptualization. Fundamentally the environmental industry is reactive and retroactive.

8.10 The energy in transportation dilemma

Transportation affects the quality of both the built environment and the global environment as much as buildings do. Our present conceptions and practices, especially regarding automobiles, are ultimately suicidal. Simultaneously automobiles shape our cities as well as the countryside. Transportation accounts for a third of all energy used in the USA and is responsible for nearly 30% of total US carbon emissions. Cars and trucks account for 63% of all petroleum consumed in the USA.

The transportation sector offers great potential for efficiency gains. The average fuel efficiency of new American cars doubled between 1973 and 1987, largely due to federal standards. New automobile models have been developed that are more than twice as fuel efficient as the average automobile today. Other gains may be possible through improved public transit, urban and regional planning and, over time, greater use of alternative fuels such as electricity and hydrogen.

Unfortunately, the public attention received by transportation is much less than is warranted by the amount of environmental damage and energy consumption it causes. Research and development of new efficiency technologies, alternative fuels and policy options to reduce vehicle miles travelled have been extremely limited. With the exception of fuel efficiency standards for automobiles – which are still well below potential levels – little has been done to reduce energy use in transportation.

8.11 The dimensions of fuel use

The need for petroleum within the USA is unrivalled in the rest of the industrialized world, as statistics show:

- the USA has only 5% of the world's population but uses 26% of the world's oil;
- the USA is more dependent on foreign oil today than during the last two Middle East crises, half the oil consumed in the USA is now imported;
- about half the US trade deficit is due to oil imports;
- the USA uses some 17 million barrels of oil a day, 43% of it in cars;
- Americans use twice as much gasoline per capita as the world's second-largest consumer, Sweden;
- more than 188 million passenger vehicles were registered in 1990, more than one per licensed driver;
- 82% of all travel was by private vehicle, public transportation was used for only 2.5% of all trips taken;
- Americans drive about 1.4 trillion miles each year – the equivalent of 7256 round-trips to the sun.

8.12 The results of fuel use

Every car that burns gasoline spews carbon dioxide from its tailpipe – 20 pounds of CO_2 for each gallon of gasoline burned. The transportation sector (primarily automobiles, trucks, buses, aircraft and trains) accounted for about one-third of the nation's CO_2 emissions in 1988 and about 8% of global CO_2 emissions – close to 1.7 billion tons.

Vehicles also emit a variety of pollutants, such as nitrogen oxides (NO_x) and reactive hydrocarbons (HC_s). Scientists refer to these pollutants as 'ozone precursors' because they react in the presence of heat and light to form ozone – 'smog' – in the lower atmosphere, or troposphere. In 1986 the transportation sector accounted for over 30% of these precursor emissions.

The relationship between global warming and ozone pollution is a vicious circle: (a) pollutants combine with heat and sunlight to make tropospheric ozone (smog); (b) smog contributes to global warming; (c) more heat means more smog.

Factors contributing to the 15% increase in petroleum use in the USA from 1982 to 1988 including the following:

- a substantial increase in trips in less efficient light-duty vehicles (pick-up trucks, mini vans and jeeps) – a 34% increase from 1980 to 1987;
- a 25% increase in total vehicle miles travelled from 1980 to 1987;
- a large increase in the number of vehicles on the road – 20% more in 1987 than in 1980;
- increased amounts of urban driving and traffic congestion, both of which have reduced miles per gallon by an estimated 15%.

This type of data is becoming increasingly available and increasingly better documented for automobile use. But environmental accountability is, apparently, superseded by the joy and freedom of driving oneself along a speedy open road. Unfortunately, the hard information for the impact of energy in buildings is less substantially documented and certainly much less disseminated. However, accountability of the value or cost of quality accommodation within buildings is already well understood, documented and appreciated.

Thus a major effort in the next decade must be to generate and make available the energy data base for buildings. For instance, we are only now beginning to get a comprehensive understanding of the total life costs of the use of aluminium in building. Beyond the cost and environmental accountability is our ability to produce sustainability comprehensively included in our collective designs of the built environment? What are the trade offs? Is aluminium a better choice for door and window frames than hardwood?

8.13 Architecture as public service

Unfortunately architecture as a practice of engagement for the benefit of society is an optional or elective service in the eyes of society. Unlike medicine or law which are need-driven and thus provide experts as remedial service in life-threatening situations, architecture and planning are usually market-driven, where the cost of service is not always equated with the value added to the product of the built environment. In essence, a reconceptualization of these terms of service is itself a necessity if the built environment is to play a role in the restoration of a balance to the natural world. Such a reconceptualization would dramatically restructure both the expectations and the design results of the built environment. It would also revalue the services of building designers.

8.14 Future practice

The future practice of architecture and planning will probably see a growing divergence between levels of service and thus between rates of remuneration. Sharper lines will be seen dividing the design professions.

There will always be those who offer to work for minimum fees and thus will have to provide nominal and minimal service. Their buildings and developments are the ones that will require remodelling, reconditioning and rebuilding, as well as the continuous use of consultants and specialists to upgrade or correct faulty original decisions. In contrast will be those individuals and collective offices that offer comprehensive services. To do this they must offer multi-tiered contracts, and build the use of specialists and experts into their services from the beginning, each representing knowledge in great depth, but of a limited area. Only in these circumstances can architects maintain their positions of leadership as master builders.

8.15 Architectural education for the 'good life'

'You are a guest of nature: so behave!' said Australian architect Glenn Murcutt to students in Phoenix, Arizona, in February 1991. But architecture and Planning as currently taught and practised are an enemy of the earth as our natural host environment. The 'good life' vision of the industrial age that is the architect's dream is a driving force of global environmental pollution. Of all the recognized ecological systems it is human urbanism that seems most destructive of its host. The worldwide quest for the consumer oriented urban life, supplied and demanded by massive industry, threatens to destroy our host environment. If we do not redirect our studies and activities towards life, then we will suffer or kill other forms of life and will possibly die by our own hand. Thus, architects and planners, as the inspired formgivers of the built environment, must create new models of practice in responsibility and completeness that begin by respecting all life.

8.16 The global mandate

Problems with energy and buildings are not unique to the USA. Everywhere architects who are serious about their professional responsibility are beginning to go beyond consciousness raising, beyond putting windows on the sunny sides of our houses. A typical mandate was given at the World Biennial of Architecture held in Sofia, Bulgaria, in July 1991:

> Our planet is deteriorating. There is a global need to save the ecological base on which all life on the planet ultimately depends. We have to achieve an organic coexistence between the natural and built environment, thereby ensuring a better life for present and future generations.
>
> Our cities are sick. The air we breathe, the water we drink, the streets we walk in are polluted. The noise all around us causes

neurosis; the traffic is at a standstill. There is little opportunity for human contact or the expression of cultural identity in the new cities and districts. They are places without life or soul.

The contemporary city does not stimulate human beings; it degrades them. First it kills Nature, and then it kills Man.

As we approach the 21st century, it is becoming imperative to develop new ideas, concepts, principles, criteria and standards for the creation of ecologically sound cities.

The improvement of the quality of life in cities through mitigation of the environmental impact of urban growth and through the efficient use of new clean technologies stands at the top of the international agenda.

We must learn to build healthy, vital cities. They must be human, ecologically sound, new technology cities – ECOPOLISES.

Every country has to find its own solution, based on its particular circumstances. However, there are underlying principles that will be common to all solutions.

As the 21st century draws near, the greatest challenge is how to create a 'non-closed', wholly open city which, like living matter, will organically coexist with the environment instead of destroying it.

There is a global need for such Ecopolises – in Africa, Asia, North and South America, Europe, Australia, the Arctic and the Antarctic.

Every Ecopolis must be planned and built to achieve an integration of the climatic, cultural, technical, industrial and other conditions of the place in which it stands.

It will of necessity accommodate local, social needs, while respecting the basic principles of a balanced approach to the city's needs and its natural environment.

8.17 Architecture as an armature of natural processes

Architectural expression must increasingly celebrate performance and alliances with natural processes. The American underground architect Malcolm Wells has published an evaluation scale that could be used to rate the environmental productivity of any built or unbuilt environment (Wells, 1976; Figure 8.6).

More recent are the European demonstrations of professional commitments that have been built: Björn Berge, a Norwegian architect, founded the GAIA group 10 years ago with a handful of architects from Norway. This has developed into a network of architects from Sweden, Denmark, Germany, the Netherlands and Ireland, for mutual exchange of information and training. Berge and the GAIA group are very consistent in their ecological architecture and have built a large number of naturally ventilated ecological houses. Björn Berge has evaluated the life cycles of

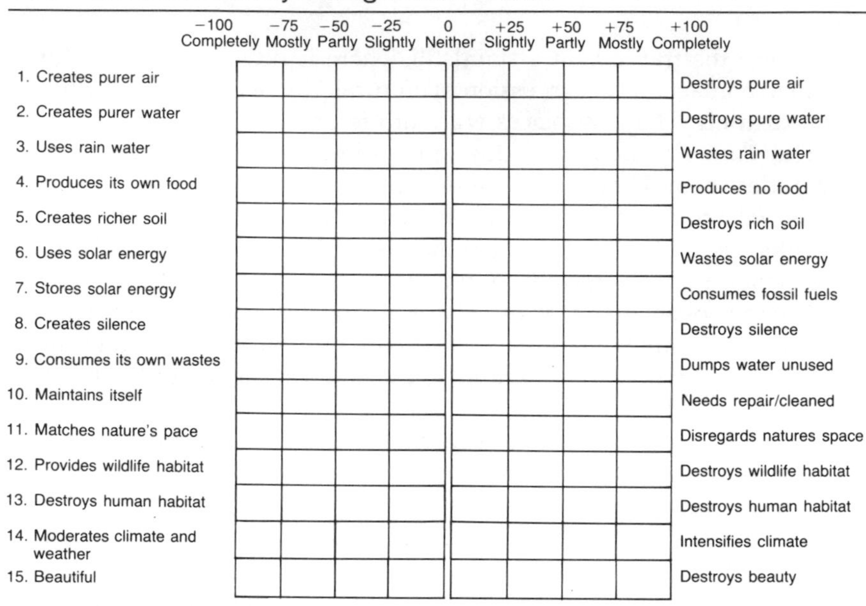

Figure 8.6

materials, the way they are produced and transported, age, etc.; his books (Berge, 1988, 1992) are thorough in their treatment of building materials and designs, their interaction with one another and the occupants.

Ton Alberts, a Dutch architect, and his partner, Max Van Hunt, have made the 2500 employee office of the NMB Bank southeast of Amsterdam an instant ecological landmark. With its inclined brick walls and its 10 sculptured towers, two of which are always oriented towards the sun, it has become one of the buildings which symbolize the Amsterdam of today. The building with its 5–7 floors gives the impression of being small in spite of its size with a budget of $US 147 million. Intimacy and a human scale have been achieved. The layout also contributes to the creation of a calm external climate. The brick building has walls which slope inwards and a shape that controls the wind. These make an exciting architectural statement and also serve as a noise barrier against the nearby busy expressway. The sound is reflected and the building is free from vibration. Ventilation in the staircase is based on the application of fundamental thermodynamic principles. Air is preheated by the sun when it is taken in through those towers which face the sun. Gardens with ponds and biodynamic cultivation have created a habitat for birds which is new for this urban environment. Three courtyards, with their biodynamic cultivation and their ponds, have also provided a habitat for unusual bird life. To be able to see a crane catch a fish from an office window is not only unique in a city, it also enhances the quality of life for the occupants of the building.

Figure 8.7 Instant ecological landmark. NMB Bank, Amsterdam.

The creative expression of humanistic, energy and ecological goals in the NMB Bank demonstrate worthy new organic architectural standards. But is the 1992 landmark Hammersmith Ark by Ralph Erskine a prototype of the future of some new organic order? Or a soft edged computer drawn London version of the old explorative speculative office building on a hitherto unbuildable degraded site? (See Figure 8.8.)

8.18 Revising architectural theories

Traditionally architectural expression has been guided by precepts such as Palladio's concern with virtue and nature. By nature is meant that buildings 'should appear an entire and well finished body. Our blessed Lord had designed the parts of our body, so that the most beautiful should be in places which are exposed to sight, and the less decent in hidden places.' Thus in his villas, utilitarian needs such as kitchens were hidden or at least subservient to a grander hierarchy in the creation of wholeness. Today the kitchen is not so unseemly. Increasingly it is recognized as the cockpit of our suburban mini-villas. Perhaps it is much too late to be shy or retiring about our essential architectural functions. We have been victimized by hiding our private but critical operative parts.

But Palladio's actual practice can be recommended much more strongly than that of many well known current architects. In his substantial single publication Palladio devoted considerable concern to bioclimate issues and to health as the basis of design decision. But these words have been ignored by both scholars and practitioners. In his built work, especially in his villas, Palladio takes his own advice seriously. Siting, orientation, building form, openings and porches have clear reference to local climate. But Palladio also dealt with such practical and utilitarian needs as grain storage, farm courtyards and the 'barchessa' or rural service buildings characteristic of the Venito. They became elements of the composition

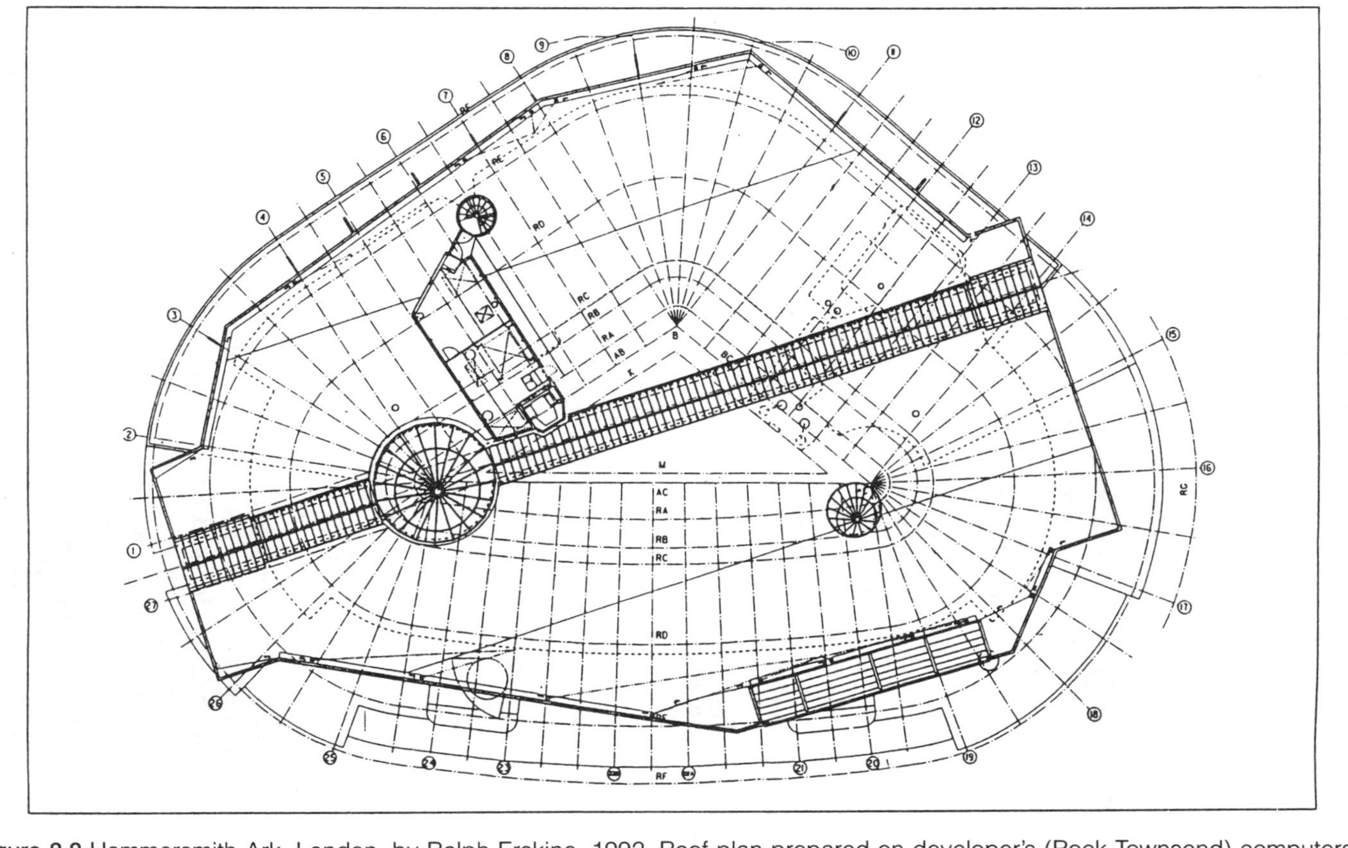

Figure 8.8 Hammersmith Ark, London, by Ralph Erskine, 1992. Roof plan prepared on developer's (Rock Townsend) computers.

reinforcing the architect's vision of an orderly as well as productive world. Aesthetics were reinforced by performance. The formal bilateral classical symmetry is visible and often blindly copied. The underlying bioclimatic order is largely invisible and is hardly noticed except for the four and a half centuries of user satisfaction.

Passive and low energy designs that use natural processes of solar heating, daylighting and space conditioning obviously represent lower impact techniques than mechanically driven, fuel supplied systems. Often the use of a small amount of parasitic power can have a large impact in controlling interior conditions. But to date this intelligence has been a kind of plastic surgery to a body which too often has all the old attributes – it is a consumerist anthropocentric machine in which a passive supply of thermal comfort is substituted in an otherwise antique model.

It is perhaps symptomatic of professional blindness that the often amazing comfort and environmental performances of vernacular architecture continue to be largely ignored by professional designers of the built environment. Often without architects, mud architecture is being rediscovered for many non-romantic reasons (see Figure 8.9).

It is perhaps also symptomatic that in the object centred and ego manic post-industrial world, the apparent failure of a rational world of

Figure 8.9 Mud architecture: Adobe Mosque and Walled Courtyard, Mali, Africa – indigenous design.

controlled absolutes is often the architectural licence for irresponsibility. Contemporary science since Einstein has found new methodologies to deal with a world too complicated to understand. In a cosmos where there are multiple truths that overlap and contradict, new science deals with multiplicity, complexity and chaos to handle the ill-defined and ambiguous. Fractal geometrics reveal whole new levels of dynamic order and system in natural and seemingly casual events. But too many architects live in a Newtonian world, fearing an exclusive determinism, a fascism of form, if obvious natural laws are followed. The complex overlapping arts of passive heating and cooling should immediately suggest parallels to the methodologies of science. Architectural opportunities of passive performance are not deterministic straitjackets. Multiple passive strategies and overlapping tactics should encourage creative and vital architectural ways within the rich menu of environmental responsibility.

There are so many passive and hybrid means of cooling and heating that need to be enlightened by the architect's design imagination (see Figure 8.10), but complex systems and high performance need not result in buildings that look like experiments in physics. A conceptual diagram does not make architecture. The documented performance of the Rocky Mountain Institute in the Alpine climate of Snowmass, Colorado, reinforces the aesthetic evidence of the joys in passive and low energy architecture. (See Figure 8.11.)

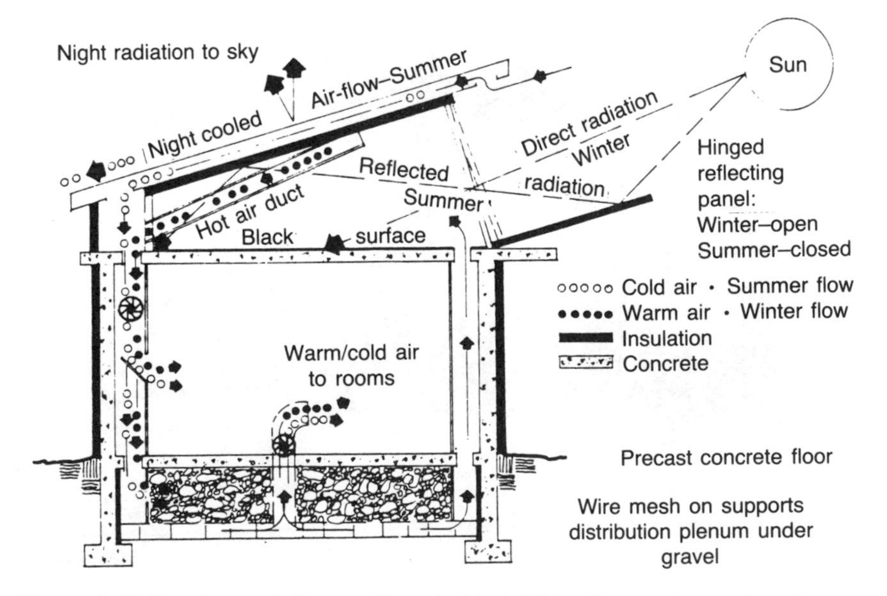

Figure 8.10 Passive radiative cooling and hybrid heating system using thermosiphon and rock bed storage by Baruch Givoni, 1977.

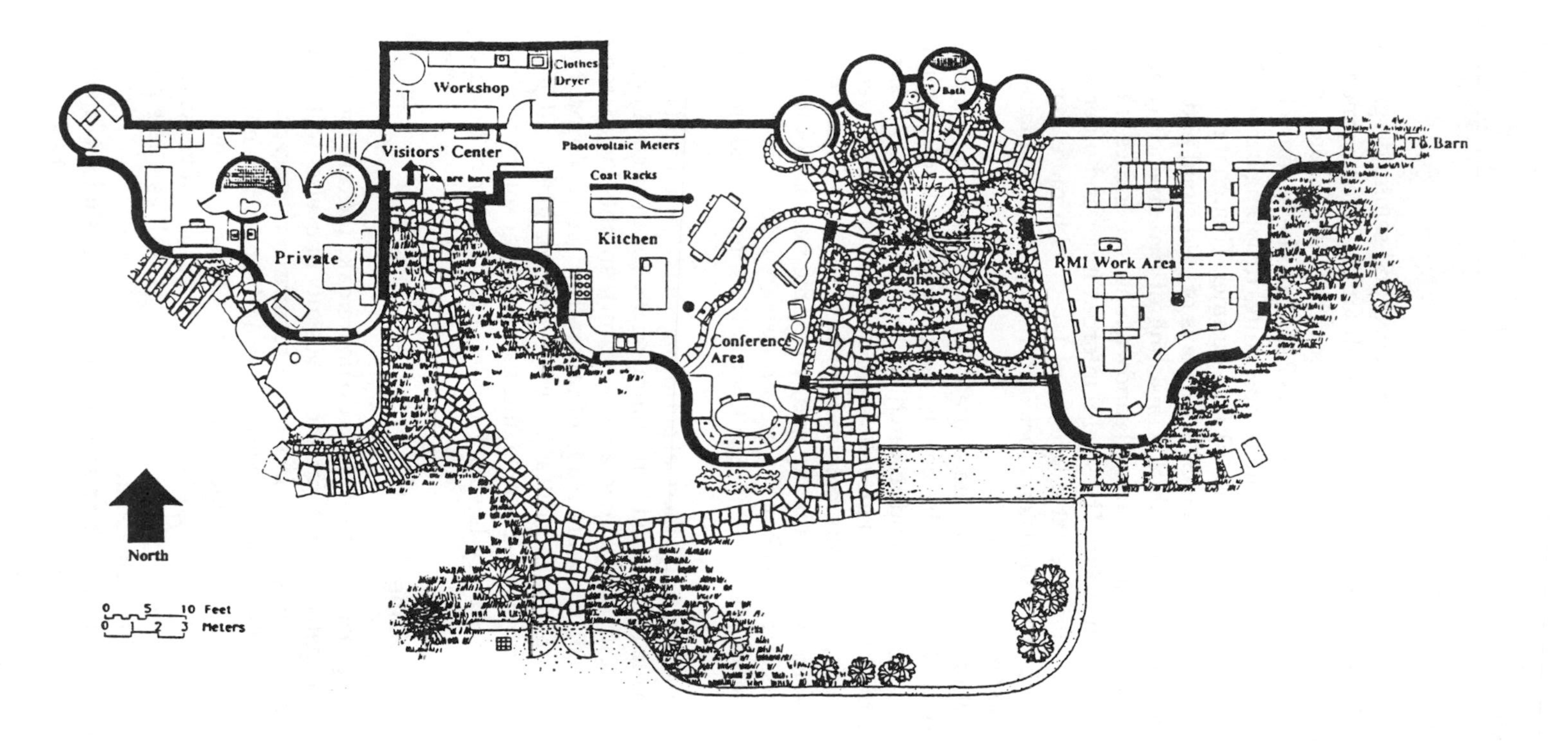

Figure 8.11 Rocky Mountain Institute Research Center and residence of Amory and Hunter Lovins, Snowmass, Colorado, by Steven Conger and The Aspen Design Group, 1986.

8.18.1. Rocky Mountain Institute

The building is passive solar, superinsulated and semiunderground. It is built back into the hill near the north lot line, then bermed the north wall and earth-sheltered the roof for esthetic and microclimatic reasons. The site's elevation is 7100 feet; the heating climate, about 8700°F days per year, with temperatures dropping below –40°F in winter and rising to the mid–80°F in summer. Though it is usually sunny, in late 1983 there were 39 solid days of cold cloud and snow.

The walls are 16 inches thick, consisting of two six-inch courses of masonry (7000+ psi mortar mix) sandwiching four inches of Freon-filled polyurethane foam. The steel-reinforced inner and outer walls were built up 20 inches at a time, and faced with more than 150 tons of thin Dakota sandstone rocks from a nearby mountain-side, within wooden 'slipforms'.

The bedroom wing is heated entirely by its south-facing windows (plus surplus heat from the people, dog, lights and solar storage tank). The living space and research center, however, also get spill over heat from the central greenhouse. Sunlight entering its vertical and 30° glazings transfers both radiant heat and warm air to the adjacent 'wings' of the building and, when the heat is not needed, out the vents in the back of the greenhouse arch. Extra heat is also stored in the arch, the greenhouse floor and contents, and the floor slabs.

Much of the building's thermal performance is due to its advanced windows, which were used here commercially for the first time. All (except the south-facing operable windows in the living room, kitchen and bedrooms) are made of argon-filled Heat Mirror. They lose only 19% as much heat as a single pane of glass, but let in three-quarters of the visible light and half the total solar energy.

(Amory Lovins, Founder and Director of Research)

8.19 Reconceptualizing the future

Architecture is proactive. If architects do not remake the case for architecture in environmental terms for the 1990s no one else will. Such a remaking must be more than a face-lift. It requires a reconceptualization of goals in the built environment. Buildings in their discrete entity are of all built objects the easiest to understand and to design as whole systems. But a radical shift in design goals for all aspects of the built environment is required. Both a more sustainable economy as well as a less polluted environment require such a shift.

A paradigm that respects and encourages all living things is an organic model recommended to support these human necessities and aspirations.

Already there are examples of architecture as well as other parts of the built environment that anticipate such a shift towards sustainability: Margrit Kennedy, at present Professor of Resource Saving Design and Construction at the school of Architecture at Hannover University, is an architect and economist. Together with her husband, Declan Kennedy, she has founded the Permaculture Institute of Europe at Steyerberg, Germany. Near the Institute they have also displayed and initiated development of permaculture principles in an eco-village. Lebensgarten at Steyerberg is also a centre for training.

Among technologies used in the buildings are those of preheating fresh air by passing it through the ground, with further heating in greenhouses, extraction of heat from the exhaust air by heat pumps and ecological composing toilets and stove. 'Grey' water is led through soil beds in the greenhouses to outdoor ponds with impervious bases. From these it is passed through pipes out into the soil where it slowly infiltrates.

There is also zoned cultivation, intensive near the houses with products in frequent use, including greenhouses and kitchen gardens and extensive further out, where apples and potatoes are grown. Near the houses there is both arable land and woodland, where the principles of permaculture and agroforestry have created an interesting ecological landscape of high yield.

Such concepts now have considerable legitimacy and predictability through the historic applications among biologists and ecologists such as the Todds of The New Alchemy Institute in East Falmouth, Massachusetts. Since 1970 their Cape Cod Ark has added to the popularization of integrated biological systems similar to the ambitious and mammoth 1990 Arizona Biosphere II. But typically architects have not been the leaders in creating ecological shelters.

An exception in the USA has been the design of nature interpretation centres. With their special educational and demonstration roles they emerged as a new and popular building type in the 1980s. They present inventive and accessible structured ecologies – often achieving great biotic richness in a modest space. Could they be prototypes of a new kind of organic architecture, especially appropriate residentially?

The New Canaan Nature Center in Connecticut (see Figure 8.12), by architect Donald Watson, with its 372 m^2 area is one of the larger examples. It goes beyond all the expectations of its nominal greenhouse shape. It is not a traditional glasshouse. Fundamentally it is a carefully articulated growing machine where thermal and luminous management have informed the architecture and its detail. Its many integrated techniques provide a variety of optimal interior environmental conditions for horticulture while being highly energy efficient. Some devices such as a combined shading curtain and insulating night shade, and the phase

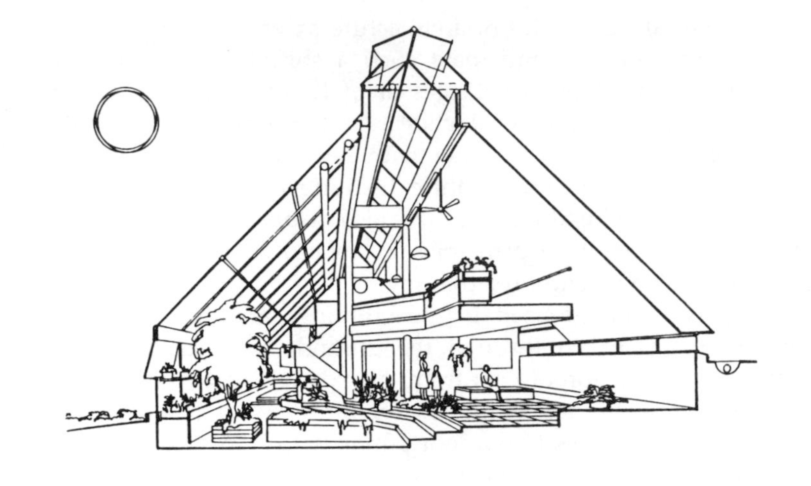

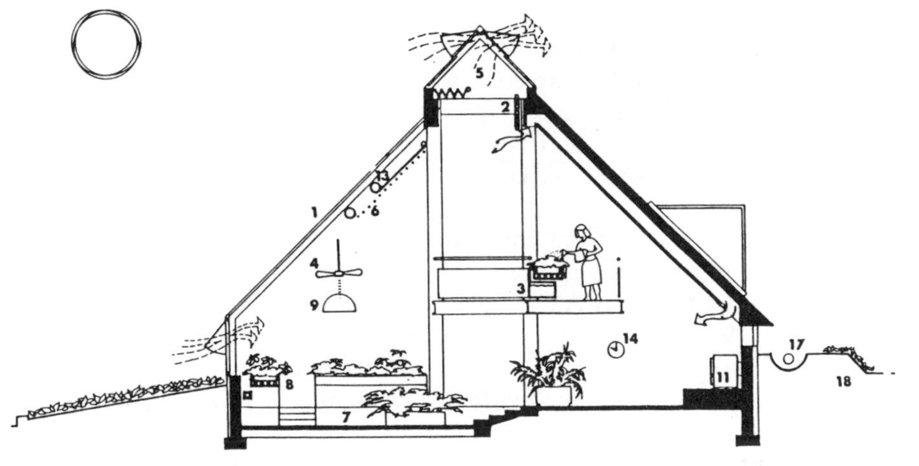

DESIGN TECHNIQUES

1 South-facing greenhouse
2 Solar collectors
3 Thermal storage elements
4 Ceiling fans
5 Roof monitor
6 Operable sun-shade
7 Earth-contact floor
8 Root-bed heating
9 Grow-lights
10 Composting bins
11 Wood stove w/heat recoverer
12 Well-insulated structure
13 Operable insulating curtain
14 Automatic temperature control
15 Energy-efficient lighting
16 Water-saving plumbing

Figure 8.12 New Canaan Nature Center, Connecticut, by Donald Watson, 1985. Cross-section of interior and section perspective.

change materials (calcium chloride) enclosed in interior partitions and railings were uniquely developed for this building. Yet the design does not overpower its contents or its purposes.

8.20 Participants in change

Perhaps finally the inability of the architectural profession to quantify the value of its service, or to prove the economics of 'design quality' could have a scientifically acceptable vehicle. Provision of passive space cooling and heating is easy to document as demonstrated by the very first fully instrumented fully passive system (Figure 8.13). Harold Hay's 'Skytherm' system in a 1967–68 continuous full size experiment in Phoenix, Arizona, under the analytic direction of John Yellott produced an easy to understand annual graphic. However, environmental and ecological productivity require new evaluation and communication models.

Within global needs to redefine the built environment, architects and

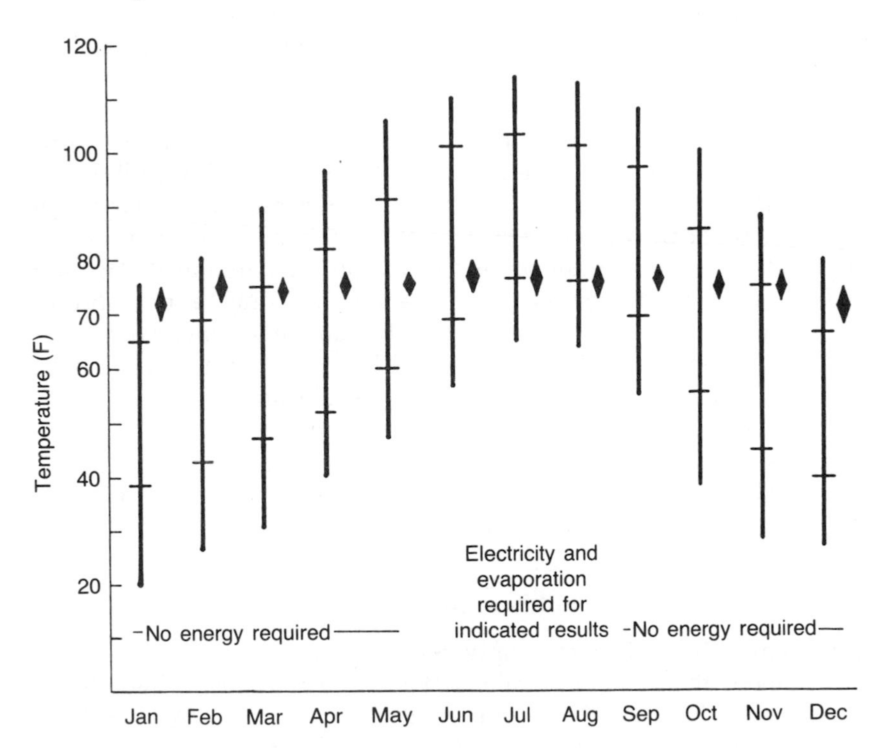

Figure 8.13 Annual passive space conditioning: 'Skytherm' performance in Phoenix, Arizona, 1967–68 by Harold Hay. Horizontal bars show monthly averages of daily maxima and minima temperatures. The diamonds indicate the average range of room temperature during a 24-hour cycle.

planners do not operate in a vacuum. Everyone who shares in the use of the built environment, whether private and domestic, or public and civic; whether it is in the infrastructure of roads, water works and waste systems, or in the workplace or home – all may participate. In addition, all users must be involved beyond the immediacy of current use. Thus revised tax laws and write downs, access to fresh air and clean water and respecting the off site impacts of every building decision must be part of our revised thinking.

It is possible to see great professional opportunity within these revisionist goals. As part of the comprehensive integrated team that does it right the first time, or as part of the correction and upgrading industry for the built environment, there are several kinds of bright concurrent futures for architects. At the least, much of the second-rate building stock already in existence must be made more viable. At a minimum architects could become house surgeons, doctors and nurses to patch up and transform the failing suburban residential fabric. In 1983 Doug Kelbaugh proposed a populist and accretive strategy. Increasing the sustainability of single family housing goes well beyond passive space conditioning (Figure 8.14). It is a future of old buildings and old neighbourhoods that have been restored, rehabilitated and reused; thus creatively recycled to new standards.

Everywhere in the industrially driven world the visible cracks of structural failure reinforce the need not just for accretive change, but for a conceptual shift to new global paradigms for human settlement; for a revision of the present belief system of habitation by consumption; and for a radical redefinition of success in the built environment. The word 'radical' means not just extreme change, not only a new original approach; but also a re-examination of fundamentals and return to the roots. It assumes a rediscovery of origins. Sustainability, not just survival, must be the key to this radical redefinition. Bioclimatic design, beyond passive cooling and heating, is one of the stepping stones. Will it have some resemblance to the familiar architectural expressions of today as illustrated by Kelbaugh in 1983, or will it be a bold statement of the new age illustrated by Steve Baer in 1971 (Figure 8.2)?

8.21 Architecture that is not benign

Architecture that is benign simply does not go far enough. In a global environment that is dramatically deteriorating because of accelerated abuses, an architecture that is only neutral cannot heal the wounds and will not restore the equilibrium of the global natural system.

Architecture should extend and enrich the natural landscape. All buildings have environmental impact – in their intervention with the existing world they modify sites, convert natural resources and throughout their

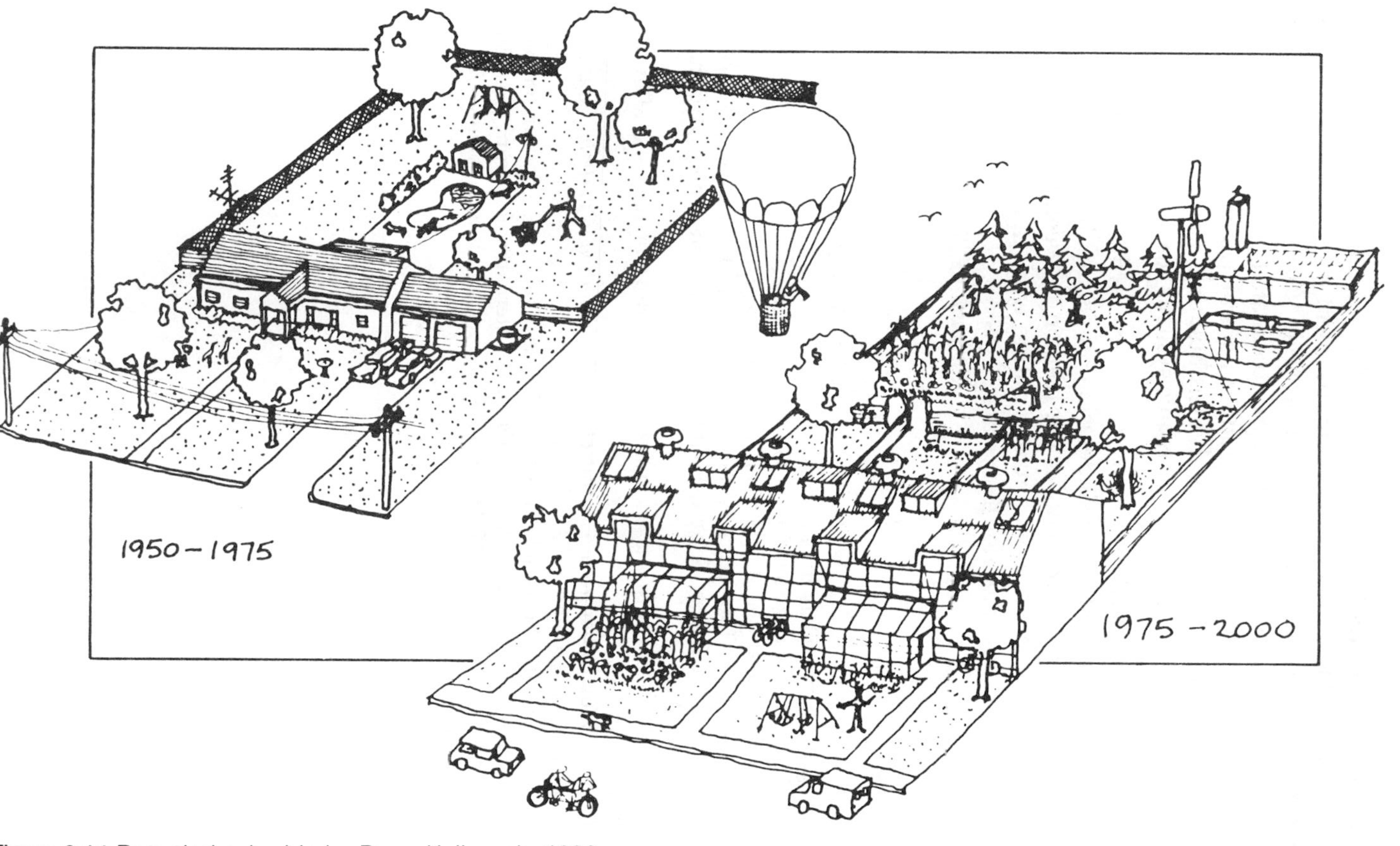

Figure 8.14 Recycled suburbia by Doug Kelbaugh, 1983.

years of use operate interactively with the surrounding environment. Seldom do buildings represent an improvement on natural processes as demonstrated by the scoring system of Wells (1976). Unlike planting a garden where vegetal productivity is accelerated, built environments provide human shelter while degrading natural water and air quality and displacing biotic liveliness with inert building materials. Can architects design those necessary new organic systems that truly support life?

> All actions take place in time by the interweaving of the forces of nature. The man lost in selfish delusion thinks that he himself is the actor. But the man who knows the relation between the forces of nature and actions see how forces of nature work upon other forces of nature and becomes not their slave.
>
> *Khrishna in Bhagavad Ghita 3.2*

References

Berge, B. (1988) *De Siste Syke Hus (The Last Sick Buildings)*, Oslo, Universitetsforlaget. 178 pp. (Only in Norwegian.)

Berge, B. (1992) *Bygningsmaterialenes Økologi (The Ecology of Building Materials)*, Oslo Universitetsforlaget. 293 pp. (Only in Norwegian.)

Kelbaugh, D. (1983) The future is not what it used to be. *Passie Solar Journal*. American Solar Energy Society, Boulder, Colorado, and Tempe, Arizona, **2**, (2) pp 99–106.

Muthesius, H. (1925) *Wie baue ich mein haus*, Verlag F. Bruckmann A.-G., München, p 82.

Wells, M. (1976) Underground architecture. *Co Evolution Quarterly*, Sausalito, California, **11**, Autumn, p 89.

World Biennial of Architecture (1991) Sophia, Bulgaria. [Invitation and preliminary application for participation in ECOPOLIS International Forum (in preparation for the 3–5 June 1992 ECOPOLIS Symposium, Rio de Janeiro, Brazil) in English and Spanish.]

Energy efficiency and the non-residential building sector

9

Deo K. Prasad

9.1 Introduction

The non-residential building sector in most countries constitutes a very significant proportion of the built environment. The central business districts, suburban centres, institutional buildings and the industrial and warehousing complexes are a major cause of concern in terms of the visual environment as well as their impact on energy requirements for their construction and operation. The resultant impact of these buildings on greenhouse gas emissions is quite significant.

The location of these centres/buildings adds a further dimension to the problem – that of transportation to and from them, to meet the service requirements they offer. The issues of urban planning and transport have been dealt with in earlier chapters [Newman (Chapter 5) and Rodger (Chapter 6)]. Industry's contribution to the global warming problem is not the subject of this chapter either. This chapter focuses largely on the commercial sector energy use and the consequent impact on global warming.

The decision to invest in a commercial property has in the past been based largely on issues other than energy or environmental appropriateness. The primary factors include affordability (company budget), prestige nature of property, location and work environment. The occupants

Global Warming and the Built Environment. Edited by Robert Samuels and Deo K. Prasad. Published in 1994 by E & FN Spon, London. ISBN 0 419 19210 7.

produce income and hence their productivity should be an issue. Since most of the operational hours are well defined and the comfort levels are also well known, it has been easy to design the mechanical systems to meet these needs. The operational cost structure of the building has long been incorporated into the operating expenses of the premises but little thought has been given to actual cost and energy management issues that could lead to significant reductions in these costs. The thinking on this is now changing as it becomes clearer that saving a dollar in expenses adds a lot more to profitability than generating an additional dollar in revenue.

The issues in non-residential property purchasing are not emotive when compared to buying a house. In the design stage as well, the structure of the design and procurement teams is often such that there is little communication. It is well known that the interaction between the architects and other consultants is mostly such that little thought is given to long-term energy use and life cycle costing. Only recently, especially since the introduction of energy standards and codes, have designers begun giving detailed thought to energy efficiency. The evolution of an environmentally benign architecture will require a much greater level of collaboration between professionals and clients. Experience in many countries has shown that both planning and design codes are required to promote the use of energy efficient and environmentally sound technologies and outcomes. The market in many cases is proving capable of adapting and demanding further improvements once the initial message gets through.

It is now clear that the benefits from energy efficient design of buildings accrues to the owner/tenant, utilities, governments and society at large. As demonstrated in the case studies later, there are actual savings to be made in operational costs of the majority of existing buildings. The reductions in electricity use have had significant effects on utility demand-side management strategies. The governments have gained not only from averting or delaying additional financing and other costs to build power generators but the reductions in their own building operational costs has implied that more money is available for other end uses. The society is the clear winner because there have been reductions already in greenhouse gas emissions with much more to come as more and more and buildings are either refurbished or built with lower energy input to achieve the same or even better quality of space and more comfort for worker performance.

In recent years, and before the onset of the current global economic recession, there has been a very high volume construction activity in most countries. The more visible of these have been in the central business district (CBD) zones of the growth cites. These have been very high-rise prestige type buildings which because of their lavish nature tend to be energy inefficient. However, the bulk of construction in terms of actual floor area has been outside the CBD in the suburbs. These have been the 2–5 storey buildings, either stand-alone or incorporated with other commercial–industrial developments. This group of buildings has been largely neglected because they are mostly developer designed and built with the express aim of low capital costs. This has implied the use of standard and pre-existing design features with overdesigned mechanical and electrical systems. The high running costs have been easily passed on to the tenants.

To make an impact on overall energy and environmental gains it would be necessary to have a rethink about not only the operational cost distribution (data on comparative energy cost indices of competing building choices should be available) but also the interaction between the design and construction teams. Energy and environmental standards and codes need to be promoted to include the most common building categories. The codes also need to be performance based to provide incentives/credits for fuel substitution. This clearly is a growing trend in most countries. The introduction of additional bonuses for on-going energy audits, addition of energy management systems and peak and total demand containment should be promoted to address a significant share of the building stock. Incentives for refurbishment of the older inefficient buildings and technologies should see a gradual changeover.

This chapter includes discussions on the energy consumption profile of commercial buildings, opportunities for energy efficiency, energy standards and codes, environmental quality and also includes a few case studies on energy saving opportunities.

9.2 Energy consumption profile

In Australia, as elsewhere in the industrialized world, more than half of building energy consumption is used for cooling, heating and lighting. For example, in the state of Victoria (Australia) in 1990, 71% of total commercial building energy use was in these three areas. This represented 28 petajoules (PJ) of energy (1 PJ = 10^{15} J). In homes, the corresponding figures were 25% and 60 PJ. The total energy, 88 PJ, was more than a third of that used by all forms of transportation, or one eighth of all energy consumption in Victoria for that year. Generating those 88 PJ released 9 megatonnes of carbon dioxide. If the trends continue, the future building energy consumption and CO_2 emissions are set to increase exponentially. Figure 9.1 shows the breakdown of commercial sector energy use in Australia.

If one looks at the total end-use of electricity in the USA it becomes clear that there is significant potential for conservation. Figure 9.2 shows that lighting is responsible for around 25% of all electricity use (1985

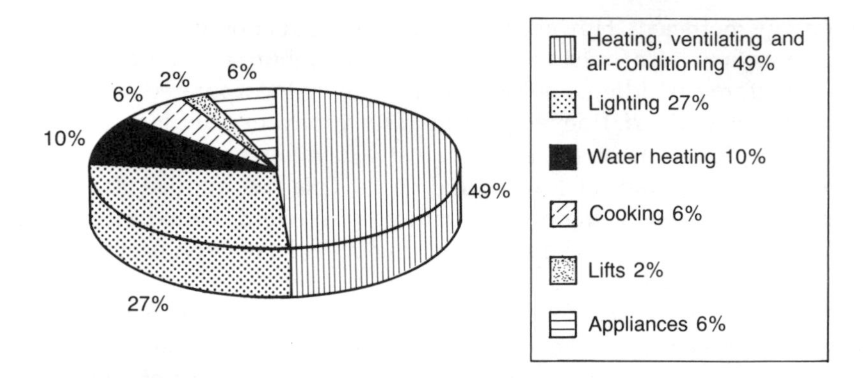

Figure 9.1 Energy use in commercial buildings in Australia.

figures). Space and water heating and cooling is also a major end-use. Commercial–institutional buildings in the USA and Europe consume far less energy today than a decade ago. This has to a large extent been brought about by the early introduction of energy codes and standards. In other countries, such as Australia, where there have not been any energy related codes (a new one is currently being developed in Australia with view to launching in 1994) there is a wide range in unit energy consumption (500–1100 $MJ/m^2/annum$). The developer built buildings are particularly inefficient because they have had little reason to be innovative and energy efficient. The fact that tenants and not developers are responsible for operational costs implies that tenants are left with a rather high consuming building. The marketing edge has not been there to bring about the change; this, however, seems to be slowly changing.

An Australian study (Deni Greene Consulting Services, 1990) looked

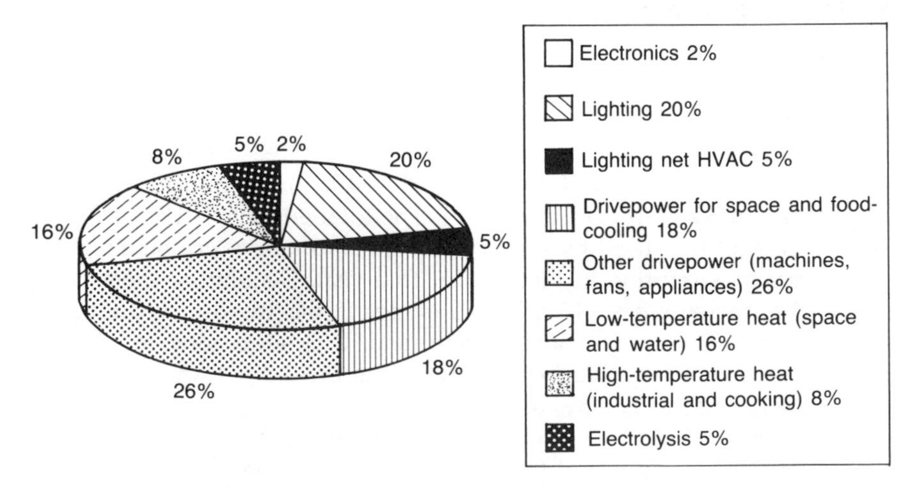

Figure 9.2 Approximate end-use of USA electricity, 1985.

at the commercial sector energy use. It studied the 1988 energy use and the resultant CO_2 emissions and compared them with a base case growth scenario to the year 2005 and another scenario of rapid implementation of energy efficiency in buildings. It was found that by the year 2005 the sectoral CO_2 emissions could be reduced to 18.8% below the 1988 levels. This highlights the potential for energy conservation in this sector in Australia. There was found to be about 60% improvement in efficiency for heating, ventilating and air-conditioning, 75% for lighting, 40% for lifts and 50% for water heating (including solar substitution and cogeneration). Figure 9.3 illustrates these findings.

9.3 Opportunities for demand side management (DSM)

Energy utilities (mostly electricity) have in recent years turned their emphasis from 'produce more and sell more' to the careful management of their demands which has led to a lot of attention being given to energy conservation. Utilities actively seek to influence the way in which consumers use electricity. This permits them to shape the demand curve and hence achieve substantial benefits by deferring costs associated with the additional supply of electricity. To encourage such conservation, utilities have designed a large number of quite novel incentive programmes which provide rebates for incorporating energy management systems, selection of equipment and compliance with energy codes. A key advantage of all this is the overall reduction in energy consumption leading to a significant drop in emission of harmful gases into the atmosphere thereby helping curtail the global warming process.

Morron (1991) grouped the common DSM programmes under six generic categories:

- Peak clipping: in which customer demand during peak hours is reduced. Supervisory load control of a customer's appliances is the most common form of peak clipping.
- Valley filling: in which demand is increased during off-peak hours, putting unused capacity to work. The result is a lower average price of electricity to the customer. A common end-use for this is thermal storage (for water and space heating).
- Load shifting: in which existing load is moved from peak hours to off-peak. Thermal storage is a common case as well.
- Strategic conservation: in which consumer energy efficiency is encouraged in selected applications to lower total systems demand. This defers the need for new capacity and may reduce the average service cost. Building weatherization and energy audits are examples of this.
- Strategic load growth: in which demand during selected seasons or particular times of the day is increased. Like valley filling this load

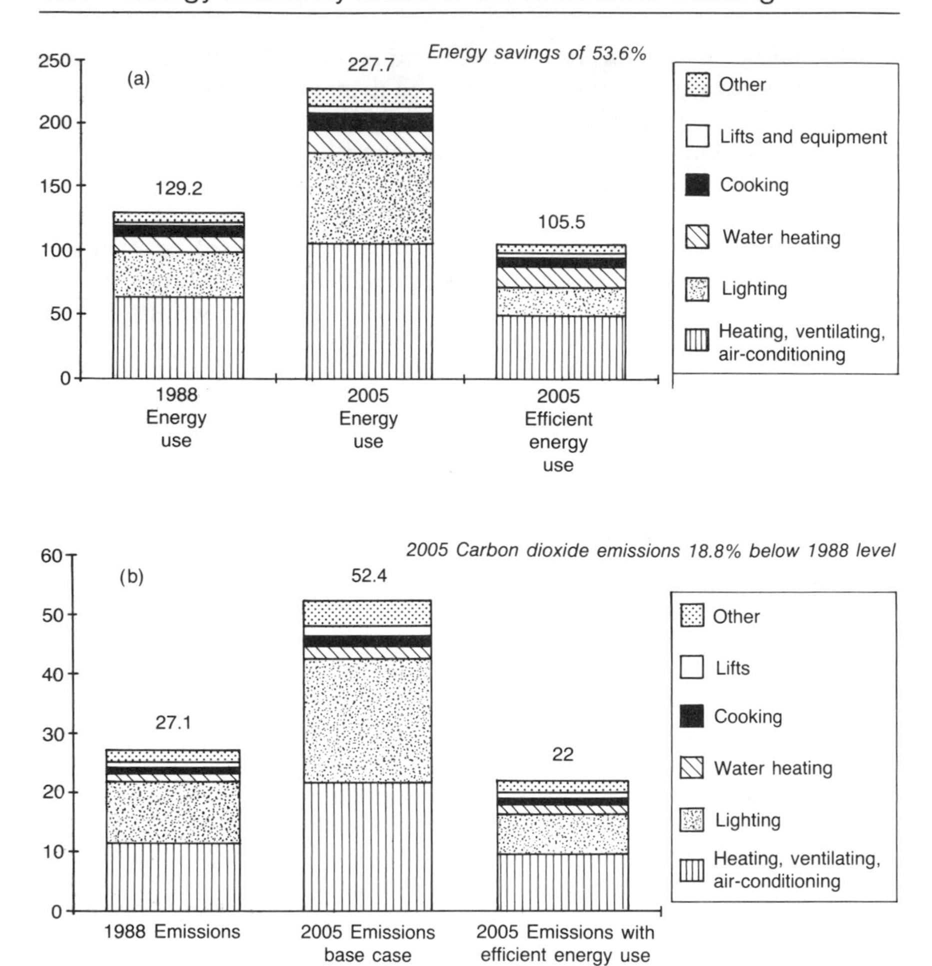

Figure 9.3 Energy saving opportunities in the commercial sector. (a) Energy saving potential (PJ); (b) carbon dioxide emissions (millions of tonnes).

increase can lower the average cost of service to consumers by spreading fixed capacity costs over a larger base of energy sales.

- Flexible load shaping: in which the load is adjusted according to operating needs, thereby gaining more flexibility in supply planning. Load shapes can be flexible if they present customers with options such as periods of interrupted service in return for various incentives.

A wide range of utility incentive programmes has been geared to promote demand management. It is, however, evident that not all schemes are directed towards energy conservation, but, on balance they do lead to an increased effort to manage resources better and to achieve greenhouse gas reductions.

9.4 Strategies for energy efficiency

9.4.1 Envelope

It is doubtful if the contemporary meaning of façade was present in the minds of those positioning the Parthenon on the Acropolis in the Hellenic era. The use of the colonnade highlighted the spiritual heart of the building inside. The hierarchy of space over external façade in western architecture continued through the medieval period. During this period, façades were formed by structural modules of Gothic space which were progressively added as construction activity continued with periods of inactivity. The Gothic style adopted in the 19th century added to the assortment of façade components.

The Italian Renaissance painters revealed their awareness of the meaning of façade in dealing with civic spaces formed by various building façades around the procession of ceremonial and everyday life. This was revealed through Palazzo façades. These classic forms were based on a rusticated base with openings which contrasted with the much larger openings on the upper floor, all accentuated by the eaves overhang which was consciously pre-positioned to the whole façade and not the functional requirements for shelter.

The contextual appropriateness of façades has been shown to vary with cultures and climatic regions. Façade fenestrations should highlight an architectural character that is region specific. A building sensitively placed on a site can embody the cultivation and transformation of an urban culture over its history. It should reveal the approach to addressing climatic problems in the region. However, influences vary with growing internationalization and flow of ideas. There has been some degree of repetition without consideration of their appropriateness. With a clearer understanding of the fabric of the façades through research, development and testing, it should be possible in the future to maintain a regional specificity while using common materials in countries far apart.

The visual significance of the façade upon a surrounding can be highlighted to some extent by comparing the relative surface areas of the façade and the surrounding street. To support adequate circulation, street areas can be about a third of the city block. However, a tower block today can represent many times more solid surface area. Hence the façades have become visually a very dominant part of the streetscape. Glass now plays a significant role in the design of all building envelopes. This has had impact on the overall energy use in such buildings.

Windows really are the weak link in the building envelope, yet today's commercial architecture uses vast walls of glass. The modern home also has far more glass than equivalent dwellings a generation ago. Unfortunately, improvements to windows have not kept pace with those in other components of the building envelope. Walls, ceilings and floors are easily insulated because they are opaque; windows present a complex challenge since they exist not only to keep the weather out, but also for light, view and aesthetics. The typical window needlessly loses heat in winter while admitting much unwanted solar energy in summer. In the hotter months we want the light but neither the heat nor the ultraviolet radiation which damages fabrics and paints.

Windows are a major factor in design because of their influence on building energy use. In the USA, about 5% of the total national energy consumption can be attributed to windows (Sullivan *et al.*, 1987). This is evenly split between windows in houses and windows in non-residential buildings. They noted that by providing daylighting, windows can also influence the 5% of the total national energy attributable to electric lighting. In the case of residential units, which are heavily envelope dependent, windows strongly influence annual heating and cooling requirements.

Recent developments in window design and technologies has permitted increased glass areas due to their tailored properties for various locations. Change from sheet and polished glass to float process has led to marked process energy efficiency. The production of heat absorbing glass resulted from concerns for energy and thermal control. These come in a range of colours: light to dark grey, bronze, green, blue, pink and amber. The same concerns have led to the development of variable transmission glass and a wide range of coatings.

Smart glass reduces space heating and cooling loads while increasing daylighting by varying its properties, thus permitting large areas to be used without negative consequences. Three key types are being developed. Thermochromic glass comprises a clear film which when heated above room temperature reflects sunlight by turning opaque white and then turning clear when cooled. The degree of opacity is also determined by the strength of solar radiation, so a window in a cool room exposed to high sunlight levels remains clear but in a warm room turns cloudy. In general these films are made of organic gels which alter their chain shape or length dramatically over short temperature spans. Photochromic glass on the other hand undergoes reversible change on illumination.

Electrochromic glass has received the greatest attention over the last decade. This contains compounds which undergo reversible colour changes when a small voltage is applied across a thin layer. They have been used in a few test cases only.

The need to maintain high transmission with low conductance has led to the development of insulated systems of various types. Two key techniques with potential are: the use of aerogels and evacuated glass units. Aerogel is an unusual type of microporous optical material that combines high transparency with excellent thermal insulating properties. A window utilizing this material can have an R-value approaching that of an insulating wall while transmission of clear glass. Similarly, evacuated glass technology relies on the resistive property of a vacuum with the panes held apart by glass bead pillars. The smart glasses and the evacuated glass have been in developmental phase for some years but their wide scale application is not foreseen before the end of the century.

The energy performance of building fenestration systems results from a complex interrelationship between glazing properties, window management strategies, building use and climatic factors. The interaction of these factors is complex and therefore complicates the selection of fenestration for optimal energy performance.

In a warm climate a window often lets in too much energy which is denoted as cooling load. Using solar control systems one can exclude the solar radiation at $0.7<\lambda<3$ μm which carries about 50% of the total solar energy. The desired optical properties are transmittance at $0.4<\lambda<0.7$ μm and high reflectance at $0.7<\lambda<3$ μm. In cold climates a window is normally a heat leak requiring insulation against heat loss while optimizing solar heat gain. A system should therefore have high transmittance at $0.3<\lambda<3$ μm and high reflectance (low emittance) at $3<\lambda<100$ μm. In a temperate climate it may be that a coating which is suitable during the winter leads to summer overheating or vice versa. Clearly there is a need for compromises and ideally for dynamic properties of glazings. This has led to the development of 'smart windows' as described.

9.4.2 HVAC systems

Heating, ventilating and air-conditioning systems are used in buildings primarily to maintain occupant health and comfort. It allows for well-regulated temperature and humidity control, air quality/quantity control and elimination of odours, etc. It is often found that such systems are over designed to an extent that they are very energy inefficient.

In order to minimize the HVAC loads it is important to address the architectural problems first. Indeed, this preventive approach ensures that the choice of building fabric, design of the envelope, the balance of thermal mass and insulation, aspect ratio, shape and size, daylighting opportunities, ventilations options and solar gain opportunities are well handled. The passive solar design principles should be investigated in detail for every building. The research efforts of the International Energy Agency Task XI Report (1992) on passive solar commercial and institutional buildings should be an important reference on opportunities for passive design.

Another key factor to recognize at the design stage is to build in options for passive operation of buildings at times when the weather is not too demanding. During the summer months when the issue of thermal comfort is of greater importance than energy considerations, the building should be able to be operated without additional energy input for cooling. Other operational features need to be understood in detail. This includes time control on air-conditioning, the distinction between human comfort criteria and other looser requirements, air handling and temperature control variances and avoiding overlapping between heating and cooling. Beyond this, space zoning and the choice of the most energy efficient systems (plant and equipment) is very important. Knowledge of the most energy efficient systems is now widely available among the design engineers.

Issues of basic housekeeping are also very important. The use of energy management systems and controls allows for optimization of energy. Intelligent systems these days are able to undertake proper resource management to meet user needs while shutting systems, or parts of it, when not in use. Task lighting controls, zone cooling/heating controls, daylight optimization and opportunistic heat recovery are examples of this. Regular energy audits and energy monitoring are a necessary part of optimizing energy use over time. These audits can provide important information which should be followed up. Factors such as thermostat settings need to be looked at for different climatic locations because if the ambient conditions are not too far from the setting, perhaps there is no need for conditioning if the building is properly designed. In Australia, there are a large number of buildings (mostly government owned) which do not have central heating–cooling systems.

Fuel substitution in buildings is now being given necessary credit in most energy codes. The use of solar and other renewable energy sources has tremendous potential for off-setting the conventional energy needs of a building. Most low temperature energy requirements can now be met cost efficiently by solar thermal means. The use of photovoltaics and other cogeneration methods also has significant potential. Other innovative methods such as waste heat recovery, also play a major role. The use of life cycle costing and internal rates of return techniques often shows very positive results when applied to these alternative energy solutions. The case studies discussed later demonstrate this. Any building design should consider the wide range of possibilities and aim to consider environmental sustainability issues as well.

9.4.3 Lighting

Lighting is clearly a large user of energy in commercial buildings. The process of the architect designing the building shell and passing the drawings to the electrical engineer for electrical layouts has in the past

Table 9.1 Advances in lighting (Source: Lovins and Sardinsky 1988)

Technology	Efficacy (lm/W)	Life (h)	Index [h(lm/W)]
Candle	0.15	7.5	1
Carbon incandescent	3	1 000	2.7×10^3
Tungsten incandescent	12	1 000	1.1×10^4
USA source mix 1988	44	12 000	4.7×10^5
USA potential 1988	595[a]	>12 000	$>6.3 \times 10^6$

[a] Equivalent efficacy including not only improved hardware to produce light and control its delivery but also improved use of that hardware by supplying light that is much easier to see with, so less is needed to achieve some visual performance, substituting natural for artificial and providing the optimal amount of light at the right time and place for task.

led to the least imaginative lighting solutions in too many buildings. The lighting designs, system selection and controls all have led to inefficiencies. One of the basic factors has been given the least regard in lighting designs. Daylight optimization not only provides most quality solutions but leads to energy efficiencies with good rates of return. Design for daylight optimization needs to be acknowledged early in the process to incorporate appropriate façade design (including window type and size decision), aspect ratio and layout, internal and external reflection controls and exposure as well as a sensitive electric design which complements the overall lighting needs. The use of some of the advanced glazing systems being developed permits greater levels of innovation in design allowing for a good balance between lighting needs and the resultant heat gain/losses. More and more designers are now looking at such innovative options.

The developments in lighting technologies have also been phenomenal (in the order of a 600-fold drop in lighting service cost). Table 9.1 summarizes performance improvements over time. In addition, several-fold improvements have been forecast before the end of this century. Lighting directly uses between 20% and 25% of USA electricity. The proportion is even greater in some other countries. Major savings in the use of electricity will lead to substantial reductions in greenhouse gas emissions especially in countries such as Australia which use coal fired stations for generating electricity.

9.5 Building energy standards

In the past reliance has been placed on designers to implement appropriate strategies for energy efficiency in buildings. This has not been followed. The reasons for this range from lack of appropriate knowledge to client pressure for lower capital costs and aesthetic appeal. It has not been understood that both these requirements can now be met while meeting the energy efficiency criteria as well. Most developed countries however, have realized this and have implemented energy efficiency standards for buildings.

It has been the belief for some that the climate of Australia is not cold enough to demand enforcement of such standards. However, the energy saving potential in buildings as shown by recent reports point to the contrary. The performance criteria for buildings in all Australian locations vary from those in USA or Europe, pointing towards the need for local standards and not merely the adopting of others.

The matter of energy conservation standards in the USA has been complicated by the geographic, legal and administrative diversity of the country. There is considerable technical uniformity, in the sense that two technically equivalent standards, the ASHRAE 90 and the Model Energy Code, exist – each with periodic updates and modifications.

Building standards in the USA are set for county or municipal level. Historically this sometimes resulted in great disparities in requirements – for example, rural areas may have no requirements reflecting local economic interests. Prior to 1970 standards were justified on the basis of public safety. Since most countries lack the resources to develop adequate standards the typical pattern was to make the Model Energy Code mandatory (similar to uniform building codes). The traditional pattern in the USA therefore has been to add local variances to the National Standard.

The ASHRAE 90–75 Standard (Energy Conservation in New Buildings) was a prescriptive document in which minimum or maximum performance levels were specified for many building components and systems. The 'envelope' section of the standard, for example, set maximum average U-values for walls and roofs, including window and skylight systems. These values varied with location and climate; decreasing in value as the climate became more severe. This part of the standard forced many designers to select double glazing instead of single glazing as reasonable window areas were allowed (note that in early 1970s 70% of all windows were still single glazed).

A second requirement for climates where cooling was a more significant problem limited the overall solar heat gain by placing restrictions on the product of shading coefficients and area, with orientation as a variable. Depending on location, either the U-value or the solar heat gain term could be the constraint. Air infiltration was addressed by referencing standard industry specifications.

A recent shift in the approach of standards has been the development of a 'performance approach'. In this case an annual energy budget is allotted to the building and the designer is free to choose any combination of building features provided the final result is less than or equal to the budget. Budgets must be set for each climatic zone and each building type. This approach requires that standard occupancy and operating conditions be defined and that recognized calculation procedures are available. In other words, alternative prescriptive packages are created. ASHRAE is now revising Standard 90 to include many of the features of these performance based approaches. For example, it should provide an energy credit for the use of daylighting. There is an increasing notion in the standards development process that the requirements for energy efficiency should not reduce environmental quality below 'acceptable levels'.

The Californian State Energy Code probably represents the most stringent set of requirements currently in force in the USA. For both residential and non-residential buildings this code provides a set of climate dependent energy budgets. There are no requirements on individual components as long as the building can be shown to meet the overall performance requirements. As an alternative to this procedure, for non-residential buildings the thermal characteristics of the envelope are specified, including the maximum overall average U-values of walls including windows. Non-residential buildings which derive more than 40% of their annual thermal energy usage from renewable energy sources automatically satisfy these requirements.

The State of Texas Building Energy Conservation Standard (1975) is divided into sections on envelope, lighting and hot water and HVAC systems. The standard is similar to the ASHRAE 90 in parts only. In the area of envelope and lighting it sets thermal performance of the shell by specifying an Energy Envelope Index (EEI) as a function of building size and location. The EEI method has received favourable appraisals since its inception. The EEI considers the influence of geometric factors, climatological variables and the interaction of internal and envelope loads.

In Europe a number of countries have set minimum and maximum performance requirements for the building envelope, lighting and other design variables. In Australia there have been a number of attempts at providing designers with suitable levels of information for energy efficient design. A number of guidelines have been produced which have fallen short of serious application.

9.6 Environmental quality in buildings

The application of energy driven standards and technologies aim to provide benefits to users, owners, utilities, governments and society at large. It is important to ensure that there is no resultant drop in quality of space. Functional efficiency and economics have previously overridden environmental quality considerations and have taken priority over the users in many instances. Studies have shown that users are now becoming increasingly aware of quality space needs not only because of concerns for global environmental responsibility but also of the relationship between worker performance and productivity and environmental quality. Documented cases on building related illnesses and sick building syndrome have highlighted concerns for environmental quality standards in buildings.

This discussion can be taken a step further and related to the overall environmental quality. This gives credence to the greening debate and the evolution of the green building. Many books are now written on green architecture, buildings or design. They all promote the concept which goes all the way from basic early decisions and commitment made at brief development stage to materials selection, procurement options and operational energy efficiencies. Concerns for increased cost of green architecture is easily being underwritten by the benefits. Owners are now becoming aware that the market demand for such spaces is on the increase and this should lead to attitudinal changes necessary for wider acceptance and investment in environmentally benign architecture.

9.7 Case studies

The rates of return on investing in energy efficiency in buildings have in many cases been found to be very positive and lead to substantial environmental benefits as well. The new building construction can easily achieve energy targets comparable to the best possible. Energy Codes in most countries insist on a minimal level of compliance. This, however, is not as easily applied to the huge stock of existing buildings which in most cases have been found to be energy guzzlers; being a product of the abundant energy mentality era.

Professional energy audits reveal very positive paybacks on investing in retrofit savings as shown below (source: EMET Consultants, Australia):

- The ventilation codes for car-parks required a Sydney car-park operating on a 24-hour basis to have the fans running all the time irrespective of occupancy. The design of a customized control system to monitor car movements and operate the various exhaust and supply air fans to ensure air quality costing A$40k resulted in an annual savings of A$198k. Further lighting savings were also possible.
- A commercial office building with hotel spaces in Sydney had an energy bill of over A$600k per annum. An energy audit revealed that an annual saving of A$100k was possible by correcting existing problems with pneumatic control systems and installing feedback loops from control spaces to reset heating and cooling and operating parameters to suit the load. Additional savings worth A$60k per annum were possible by rationalizing the heating and hot water services and associated pumping and reticulation systems. An investment of A$90k led to a saving of A$100k per annum.

These are just two examples of potential savings with others far exceeding these. Client awareness and motivation is often the key barrier to such efficiencies. Most Building Owners and Managers Associations (BOMA) have details on such retrofits in those countries. The payback for different initiatives vary depending on the scale of the problem, climate, user profile and the original design. Every environmentally responsible property portfolio holder should institute audit and energy management procedures to achieve maximum benefits through energy conservation.

9.8 Future directions

It is evident that global warming is happening; its rate and time frame for perceptible climatic and other changes may still be argued for some time. What is also evident is that our habits, expectations and the way we have built our settlements over time has had a lack of regard for the environment. The economic booms have led to high planning and construction activities with minimal thought being given to even medium-term effect of the decisions. Cook in Chapter 8 raised the issues of going back to the basics approach. The indigenous solutions were very sensitive to sustainability. The modern planners of the last few decades seem to have forgotten the key principles.

The approach to the future would need to take into account both the curative and preventive methodologies. This implies improving the existing mess and ensuring that reasonable controls are placed to build in necessary efficiencies at all levels. Broader urban and regional planning would need to be based on sustainability principles. The relationship between home, work and transport systems all need to be reconsidered. Governments have to take a proactive role in decision making to facilitate sustainable planning goals.

As noted by Professor Ehrlich in the Foreword, the impact on the environment is directly proportional to the population, levels of affluence and the levels of technological improvement. Therefore measures that control the three key factors would contribute towards impact reduction. The issues are not simple but very intertwined and require all levels of participation.

The direct and downstream pollution effects of residential, commercial and industrial buildings needs to be considered. Energy conservation in all three areas is vital to make headway. This can firstly be achieved by passive strategies to minimize energy loads. The difference in most cases can be picked up by renewable resources. The use of appropriate materials and procurement methods allows for a greener solution. Then the reduction in energy consumption by proper management systems ensures a sustained low consumption over time.

Statutory codes and standards have played a positive role in implementing energy efficiencies. The pressure should be maintained to increase greater penetration to all sectors. The policing of pollution and other guidelines is also important to maintain the momentum. The incentives for meeting energy and environmental guidelines are important to gain wider user participation. Eventually, its the wider community that needs to take a proactive role and demand for environmentally sustainable and responsible solutions. The economists and industry will follow by accounting for the environmental impact in the pricing structure, thereby leaving an even playing field.

References

Deni Greene Consulting Services (1990) A Greenhouse Energy Strategy: Sustainable Energy Development for Australia, DASSETT Report, Canberra.

International Energy Agency Task XI Report (1992) Passive Solar Commercial and Institutional Buildings, Heating and Cooling Program.

Lovins, A. and Sardinsky, R. (1988) *The State of the Art: Lighting* (March), p 348. Rocky Mountain Institute, Colorado, USA.

Morron, T. D. (1991) Demand-side management programs, *ASHRAE J*, May, 34–41.

Norgard, J. (1991) Energy for personal comfort – efficiency options and limitations, ESETT Conference, Oct 1991, Milano.

Pupilli, S. (1991) Trends in Building Energy Management, in *Future Energy Management*, Department of Primary Industries and Energy, Australian Government Publications Service, Canberra, pp 29–47.

Sullivan, R., Arasteh, D., Sweitzer, G., Johnson, R. and Selkowitz, S. (1987) The influence of glazing selection on commercial building energy performance in hot humid climates, ASHRAE Conference on Air-conditioning, Singapore, 1987.

Principles of energy efficient residential design

10

John A. Ballinger and Deborah Cassell

10.1 Introduction

Housing is distinguished from other forms of building by the following characteristics:

- No matter how rudimentary it may be, shelter is needed by everyone (other buildings are 'optional' in that needs are driven by commercial practices, technology, culture, etc.).
- Its creation is largely in the hands of the occupants or builders rather than architects. Even in countries such as Australia, the percentage of housing designed by architects is very low; it is thought to be less than 5%.
- Housing selection for most is based on emotive, very personal decision processes subject to individual tastes and preferences, as well as economic, social and cultural factors.
- Even where choice is not available, due to economic or other circumstances, the way in which the house is used is subject to huge variation. Energy consumption for the same house can differ markedly between different occupants.
- The main opportunities for reductions in energy consumption in the housing area are in the area of passive solar design, appliance energy efficiency and user attitudes to appliance use and building operation

Global Warming and the Built Environment. Edited by Robert Samuels and Deo K. Prasad. Published in 1994 by E & FN Spon, London. ISBN 0 419 19210 7.

(mainly management of windows and other openings) to control solar energy penetration and ventilation.
- All dwelling occupants have the opportunity to contribute towards reducing the greenhouse effect through their use of housing. It offers the chance of change at the 'grass roots' level, where individuals can become involved in actively pursuing the goals of energy efficiency. The greatest energy consumers, the developed countries, have the greatest opportunity to significantly reduce the amount of energy they consume.

These factors suggest that strategies for promoting and achieving energy efficiency in the residential sector need to be very different from those which are suitable for other building types.

In less materially developed societies traditional housing consumed little energy and was generally responsive to the environment. In his book *Architecture without Architects* Bernard Rudofsky (1964) examines what it is that appeals to many people seeking to find a more fundamental response to the need for shelter. This was after all the single requirement which housing originally had to satisfy. However, in the more developed countries the occupants of houses demand a high degree of thermal comfort. This is often achieved not through the design of the house but through the provision of energy-consuming space heating and cooling, as this is the less capital intensive option. Using the form and fabric of the house itself to provide thermal comfort is almost a lost art, but one which must be regained if we are to take control of our energy consumption at this level.

Although espoused by many, a simple return to vernacular housing styles is not necessarily the answer; dwellings today must respond to a far more complex set of circumstances than they have in the past. Increasing population, changing technologies, the provision of energy consuming services to houses and rising standards of living are amongst the many constraints which influence today's housing. In the developed countries the appropriation of housing as a task of specialists (whether architects or speculative builders and developers) and the demise of sensitive housing generated by the community itself has been seen. What is needed now, therefore, is education of the designers and users of houses, so that the community not only learns to value housing which is energy efficient but also understands how to achieve it.

Much of the work undertaken in the field of energy efficient housing design focuses on a form of housing typical of developed countries: in Australia, the three bedroom brick veneer bungalow would be the obvious example. Since it is in these countries that the greatest consumption of energy per head occurs, there is an indisputable logic in addressing energy efficiency within this context.

Whilst individual energy consumption may be relatively low in developing countries, population numbers are high and so a large proportion of their energy is used in the domestic sector. The United Nations estimates that domestic consumption is responsible for 68% of the energy used in rural areas and 45% of that used in urban areas, and that a very important proportion of this energy is used to achieve bioclimatic comfort by heating and cooling dwellings. If comfort is to be provided more efficiently in these countries, designers must address their very different context. Whilst certain principles of energy efficiency are universally applicable, specific responses can be expected to vary significantly.

The search for innovative and appropriate solutions may, however, be hampered by the aspirations of the potential occupants to live in housing which is patterned on that of the developed countries. Conventional forms of housing are often perceived as particularly desirable by many of those who do not ordinarily have access to such housing. This phenomenon is demonstrated by the extensive housing developments that can be found in Asia (Bangok, Singapore and Malaysia have increasing numbers of houses that would not be out of place in Europe). Such housing is neither appropriate to their lifestyles nor for the climate. A similar desire for housing which in its form represents a certain quality – in this case comfortable familiarity – can be seen in the housing of hot tropical Darwin, where people who have come from the cooler temperate climates of the southern states prefer to live in brick bungalows entirely unsuited to the tropical climate and lifestyle. Such limitations must be seen as further design constraints which demand an imaginative response.

Part of the solution will no doubt be found in the developed countries changing their own perceptions of the conventional house. As energy efficiency is placed higher on their agenda expectations will change accordingly. The fostering of a regional architecture appropriate to the different climates of developed countries (both Australia and the USA have a wide range of climates within their borders) will assist in the creation of patterns which it may be appropriate for the developing countries to emulate. The development of prototypes will play an important role in this, as will the establishment of regulations and standards which will influence the context and form of our housing.

The key to the design of housing which uses as little energy as possible to provide reasonable levels of thermal comfort is both climate and user satisfaction and/or flexibility. The climatically appropriate house is one which responds to seasonal demands, modifying the negative effects of the climate and taking advantage of the positive ones to enhance the comfort of the occupants. Houses which do this well may also have many other positive qualities: the relationship with the outdoors, the play of light and shade, the exclusion or admission of the sun, the opening up to cooling breezes or the creation of winter sun traps are amongst the possible attributes of a responsive house, all of which can add a measure of delight to the occupant's experience of the home.

Climates vary from the hostile to the benign, with some countries covering such large areas that several different climate types and severities occur in the one country. For the purpose of building design the world's climate types may be broadly categorized as cold, temperate, hot-dry and warm-humid. Within each of these regional and local variations will occur. A specific site will have its own microclimate which must be analysed in detail before any design work is commenced.

A major consideration for any climate is, which season predominates for design purposes? Winter-driven climates are characterized by a greater need for heating in winter than cooling in summer; summer-driven climates are the reverse. In some regions summer and winter needs are of equal importance. Selecting the more significant season can be complicated by the purpose of design: is it to reduce energy consumption or to enhance comfort, two aims which may not in fact require the same response. Furthermore, strategies for winter-driven climates focus on maximizing solar gain in winter; for summer-driven climates, prevention of solar gain in summer is paramount. Solar water heating needs must still be taken into account in the latter case, since summer-driven climates are often those where there is the greatest opportunity for this method of water heating.

Climate should be one of the determining factors in regional and urban planning. Ideally such planning will also take into account the many other factors which influence energy efficiency, such as land use/urban mix/proximity of work to home. This discussion will commence at the level of the subdivision as it relates to building design.

10.2 Subdivisions

It is important that subdivisions provide the potential for the creation of energy efficient housing. In this respect, solar access is the most significant factor. The disposition of streets, lot size, slope and orientation will all influence the suitability of a block for an energy efficient house. Lower density housing offers easiest solar access; with increasing densities the protection of solar access for every dwelling becomes more difficult and more care must be exercised in the form of the subdivision to ensure solar access is generally available.

The most critical stage in the design process is that of site analysis. Undertaken in a thorough manner, this process can ensure that the most appropriate area will be selected for the different elements of a housing development. An understanding of the site can also provide inspiration for the form of the development and prevent unintentional interference with the local ecology.

The collection of data is crucial to site analysis. Information must be gathered on local climate, topography, geology, drainage patterns, vegetation, ecology and existing services and structures. Climatic factors such as temperature, humidity, solar radiation patterns, wind movement, precipitation patterns (including fog or frost pockets) and exposed or protected areas on site must all be taken into account. Topography is particularly significant for the use of solar energy, since the extent and direction of shadows are affected by the steepness and orientation of a slope. In the southern hemisphere, north-facing slopes receive the greatest amount of solar radiation and are therefore most appropriate for medium density dwellings.

Various methods of data analysis and depiction are available. A widely used process entails the use of overlays, whereby base plans showing different characteristics are superimposed to obtain composite plans indicating areas of greatest and least suitability for different purposes. A weighting can be applied to those characteristics which are significant for energy efficiency to ensure that this is given due influence.

The development of a structure plan for the overall site, locating major roads, open space, facilities and areas of housing of different densities can then be undertaken. It is in the detailed project plan, however, that the greatest care must be taken to ensure that individual residential sites have adequate solar access, protection from adverse winds and access to desirable breezes.

Although whole site solar access is desirable for outdoor garden use in a temperate climate, it requires large land areas and is inherently inefficient. South-wall (or, in the southern hemisphere, north-wall) access, where the southern (or northern) façade and roof of the building is protected from shadows in mid-winter, is a more realistic goal. Individual lots can be checked for the extent of solar access available on site by the use of techniques such as the solar envelope or shadow masks used in conjunction with the solar access butterfly. Potential dwelling locations can be identified, taking into account the orientation of the lot towards the street and any development controls, such as building setbacks. Often it may be desirable to vary setbacks to achieve enhanced solar access. Other controls such as height restrictions and lot dimensions will have an effect on solar access. Planners need to be aware of the implications of all such decisions for the energy efficiency of residential developments and to do their utmost to create subdivisions which promote rather than prevent energy efficiency.

The integration of vegetation into the development should also be considered at planning stage. Too often sites are cleared without regard for the benefit which could have been gained from the retention of the existing vegetation. Selective removal should be pursued only after the completion of the site analysis and preparation of the project plan.

Additional planting should be sympathetic with the native vegetation and should enhance positive qualities of the microclimate, such as directing cooling summer breezes across housing areas and providing protection from harsh winter winds. Trees close to intended home sites may need to be deciduous if they would otherwise block winter sun. Where native trees are generally evergreen, such as in Australia, those who prefer a native theme may have to exercise greater care in the placement of trees, taking into account their potential size at maturity.

10.3 The individual site

The design process for the individual site is similar to that for the subdivision but with a much sharper focus on specific detail. As the scale reduces from the large rural site through the suburban to the inner urban, where building work may be confined to modifying an existing attached dwelling, so the significance of small variations in prevailing conditions increases.

The site analysis which must be carried out for the individual site will take into account the same factors as that undertaken for the purposes of subdivision. However, certain parameters may already be defined, such as lot dimensions, the relationship of the site to the street, location of existing services and development controls. In addition, more complex factors come into play, such as the clients' specific needs and desires, the character of existing streetscapes and the presence of adjoining properties, which may interfere with solar access to the site or which themselves may have solar access which is required to be preserved. Ideally, a site will be analysed before it is acquired, so that its suitability for the proposed form of development can be assessed. Much energy is wasted in trying to impose fixed preconceptions regarding building form and size on inappropriate sites. In many countries the desire to build conventional or traditional construction houses on inappropriate sites (very steep slopes of natural forest, etc.) has been responsible for the destruction of coastal rain forests, ecologically sensitive hillsides and the loss of the very qualities for which the area was desirable, and all at great expense.

In terms of energy efficiency, the most critical factor, regardless of climate type, is the apparent movement of the sun. Seasonal climate variations, which become increasingly pronounced as we move from the equator, must be accounted for. Closer to the equator this variation is not as great, but it should be remembered that in the height of summer the sun actually passes to the opposite side from that which is usual, i.e. in the northern hemisphere the sun in summer passes to the north, whereas for the southern hemisphere it passes to the south.

In the hotter climates close to the equator the major concern is shading and ventilation. Further away, in cooler climates, solar gain in winter is essential and the design must focus on this.

Where site constraints preclude ideal position with regard to climatic factors other strategies must be employed. Landscaping provides one of the most effective and least expensive means to modify the microclimate.

10.3.1 Landscaping

Landform, vegetation and water are essential elements of the habitat of all living creatures: each animal is dependent on a particular combination of these elements to survive. Otherwise hostile environments can be transformed by the presence of vegetation and water, the most dramatic illustration of this being found in the desert oasis. When humans first created their own fixed form of shelter they also began to modify the surroundings, and in so doing found that the effects of the climate within the immediate environs of the dwelling could be altered. Many of the great gardens of the world are treasured not only for their aesthetic qualities but also for the way in which they provide a more benign microclimate than that which is found in their often harsher surroundings.

Unfortunately modern architecture has become divorced from landscape design and the two disciplines are often seen as separate rather than interdependent. Yet the knowledge of how a site's climatic conditions may be altered by landscaping is essential if the opportunities for energy efficient design are to be maximized.

Landscaping can be used for both sun and wind control. In cool and temperate climates deciduous trees can provide shade in summer yet allow the sun's rays to penetrate in winter. In warm humid climates, tall trees with a high canopy can shade the building whilst allowing cooling breezes to pass unhindered at lower levels. In hot arid climates breezes passing through shady, moist vegetation can provide a form of evaporative cooling.

Trees may also be used to create effective shelter belts, breaking down the force of the wind and redirecting it over the top of houses built on their leeward side. Breezes can be funnelled into an area or winds deflected around areas by well-designed planting.

Because vegetation grows and changes, its effects over time will not be entirely predictable. Careful thought should be given in the selection of species to height and spread, spacing, shape or form, density and likely growth rate.

Water may be used to great advantage in hot dry climates as a means of promoting evaporative cooling. Air passing across fountains or pools can be directed into the building to cool the occupants. In warm humid climates the presence of water may have an adverse effect since the moisture content of the air is already uncomfortably high.

The land itself may be shaped to modify the microclimate, although this should only be necessary on a small scale if site selection has been consistent with intended use. Mounds or earth berms may provide shelter and be useful in controlling air movement. Earth integrated buildings are a natural extension of this idea, allowing a direct coupling of the building with the ground to make use of its thermal storage capacity.

Human-made elements of the landscape may also influence microclimate. Fences, walls and screens, which assist in sun or wind control, are obvious examples. Materials used for finishing ground surfaces are also important. Hard paved surfaces are responsible both for reflecting and for absorbing and reradiating heat from the sun. The temperature in the vicinity of such a surface may be considerably hotter than that above grass. Where large paved areas are required, planting may assist in controlling temperatures by shading the paving from direct radiation from the sun.

Regardless of how the landscape contributes to energy efficiency it is important to realize that the design of the landscape must be integrated with that of the building and not be treated as an afterthought, nor as something that may be dispensed with if the budget comes under pressure. It is sadly the fate of landscaping to be seen as an optional extra in many projects, rather than an essential part of the design. Recognition of its role in energy efficiency may go some way towards remedying that problem.

10.4 Building design

One of the greatest obstacles to the popular acceptance of energy efficiency as a normal prerequisite for a satisfactory home is the perception of such housing as necessarily strange in appearance. In the 1970s solar houses and, more particularly, autonomous houses, were used to illustrate the need for a radical change in attitudes and lifestyles, but in the pursuit of this goal an imagery was projected which most people found incomprehensible.

By the 1980s the acceptance of energy efficiency by even the more conservative bodies as a worthy goal saw the development of rating schemes which could be applied to conventional housing forms. Unfortunately such schemes were too often based on rigid parameters which failed to recognize the diversity of thermally appropriate solutions available to the creative designer.

Fortunately there are architects who have sought to find a middle ground, where their houses are not only climatically appropriate but also relate well to their sites, are functional, look good and bring joy into the lives of the occupants. Many of those architects who manage to achieve all of this are amongst the first to admit that their houses are not designed with energy efficiency as an overriding goal, nor are their designs based on careful calculation. Rather they rely on a combination of rules of thumb and previous experience of successful strategies.

For those people who do give energy efficiency a top priority, guidance is available from the many books which have proliferated since the early seventies. Some seek to provide the scientific information for painstaking calculations to predict performance, others content themselves with simple rules of thumb which are widely accepted; still others are based on computer simulations which allow them to give prescriptive rules for design.

It has been suggested that many rules of thumb are rigid, incorrectly interpreted from scientific phenomena and presented in isolation from other variables. Alternative solutions which do not obey the rules, yet achieve similar or better performance, remain unrecognized. There is undoubtedly truth in this and it is one of the tasks of researchers and writers in this area to seek methods which will enable design decisions to be made which take into account the many variables at work. The interrelationship of factors such as glass area, orientation and shading must be acknowledged if genuinely effective solutions are to be found and designers are to be given the freedom which they need to create varied and imaginative buildings.

For a design to achieve the aim of energy efficiency it is first necessary to know when and why energy is consumed in a house. It is not enough to simply know that more than one third of domestic energy consumption, say, is used to meet space heating and cooling needs. The user's comfort preferences and the forecast occupancy patterns for the dwelling must be identified before specific user requirements for heating and cooling can be identified.

It may be necessary to generalize about comfort standards when designing for an unknown client. Decisions must be made about acceptable conditions in order to establish thermal design criteria. Extensive research has been carried out into the conditions which are agreed by a majority of people as supporting thermal comfort. Major variables which affect thermal comfort are air temperature, mean radiant temperature, air movement and humidity as well as level of activity and amount of clothing worn. Degree of acclimatization and individual tolerance levels and attitudes are also significant. Central to the question of thermal comfort design criteria is an awareness of the prevailing climatic conditions, and in particular the relative significance of heating and cooling needs.

The key to energy efficiency for thermal comfort is the control of heat gain and loss. Heat is transferred by conduction, convection and radiation. During the underheated season, radiant solar gain should be promoted and conductive losses through the building envelope as well as convective losses via cold air infiltration and warm air exfiltration should be resisted. The usefulness of solar gain as a heating strategy will depend on the severity of the climate and the amount of solar radiation available at critical times of the year. Passive solar design, which incorporates a means of collecting, storing and redistributing solar heat, can reduce dependence on auxiliary heating methods wherever there is solar access combined with an availability of solar radiation.

During the overheated season, conductive gains through the building envelope, convective gains via infiltration of warm air and radiant solar gains should be resisted. On the other hand, conductive losses through the building envelope by earth-coupling, convective losses by ventilation and radiant cooling should be promoted. Evaporative cooling is a further method of modifying conditions where humidity is not high.

Major design elements of the house which influence heat gain and loss are orientation, glazing, thermal mass, insulation and ventilation. Each of these elements must be carefully manipulated so that their combined effect produces the most appropriate response to the particular climate in which the housing is to be located.

In cold climates it is generally accepted that some form of active heating system will be required, although the literature reports of many cases where the energy needed was negligible. The main aim must therefore be to minimize heat loss, although if winter sun is available then windows should be oriented towards the equator. Double glazing is essential to avoid a net heat loss.

The thermal mass of the building is not significant unless the heating is to be intermittent. For many family homes heating will be run continuously throughout the winter and so, once the building is warm, thermal mass has little effect on internal temperatures. However, if heating is periodically turned off for a long enough period to allow the building to cool down then too much thermal mass can slow down the internal temperature rise once the heating is restored, as much of the energy will go into reheating the thermal mass. A well-insulated lightweight building may be more appropriate in this case.

In either case, the building must be fully insulated with additional precautions such as the insulation of the perimeter of concrete slabs on ground. Careful attention must be paid to weather-sealing the building to avoid the loss of warm air through exfiltration and the infiltration of cold air. Minimum ventilation levels should be kept to below 0.5 air changes per hour, which is generally considered to be a minimum requirement for health. Air locks should be provided at entrances. Protection of the building from winds will also reduce heat loss. In very severe climates where even lower ventilation rates are desirable it may be necessary to use air-to-air heat exchangers and other methods to ensure that healthy living conditions are maintained.

In temperate climates prevention of heat loss in winter is also critical, although measures need not be so stringent and there is generally the opportunity to provide most of the heating needs of the building by direct solar gain. It is in this climate that the passive solar house has its greatest application. General solar design principles, such as orientation of the windows of living areas towards the equator, coupled with sufficient thermal mass to store the heat collected and allow it to be reradiated at night, must be followed.

Windows or other glazing must be provided with shading which allows low-angle winter sun to penetrate yet excludes high-angle summer sun. Such shading should prevent overheating in summer. Adequate ventilation should ensure summer comfort on all but the hottest days.

In hot dry climates prevention of unwanted heat gain is the overriding consideration. Very high daytime temperatures contrast with low night-time temperatures (i.e. the diurnal range is large), so the three most important factors are thermal mass, roof insulation and shading of walls and windows.

A large thermal mass acts as a heat sink during the day and releases heat at night, thus evening out temperature variations and maintaining a more comfortable internal temperature. Glazing must be limited in area to prevent heat gain by conduction and must be fully shaded to prevent solar gain. Lightweight elements such as roofs should be well-insulated. Double-skinned roofs with ventilation provided between the two surfaces are appropriate. The lower roof in such a design should be insulated with Reflective Foil Laminate (RFL).

Light-coloured external surfaces can assist in heat rejection, not only by reflection during the day but by emittance at night. Radiation to the clear night sky is an important factor in heat dissipation in hot dry climates.

Traditionally, courtyards have provided a means of turning away from the hostile environment and focusing on a sheltered area where water has often been included as a means of evaporative cooling, ensuring an air supply that is cooler and less dusty than that outside. Roofs sloping towards the courtyard may also assist in draining cooler night air into the enclosed space.

In warm humid climates there is little diurnal variation and so thermal mass cannot be exploited in the same way. During the wet season (usually the most uncomfortable time of the year) sky conditions also tend to be cloudy, so night sky radiation is not available as a means of cooling. High humidity also makes evaporative cooling inappropriate.

Adequate ventilation is essential to prevent the internal temperature rising above that of the outdoors and to encourage air movement which has a cooling effect on occupants. Long narrow houses only one room deep allow maximum cross-ventilation and, if correctly oriented, also minimize exposure to the sun on the east and west. In areas near the coast elevated houses may catch cooling sea breezes better than those at ground level and the sheltered space under the house is useful for activities during the wet season. In inland areas slab on ground construction may be effective as a form of heat sink. Separate sleeping areas using lightweight construction with little thermal mass may be desirable for summer nights.

Generally, a lightweight building which cools down rapidly with any drop in external temperatures is desirable. Walls should be shaded by verandahs, roof overhangs or planting. The roof, which is the element most difficult to protect from the sun, should be given special attention so that solar gain is not transferred to the ceiling, raising the mean radiant temperature internally. Strategies include using a reflective roof surface, using reflective surfaces for the underside of the roof and the top of the ceiling, using reflective foil under the roof, having a separate ceiling and ventilating the roof space.

A further consideration regardless of climate type is the amount of daylight in a home. Most people express a distinct preference for lighting to be natural rather than artificial, therefore ample daylight should be provided for this reason as much as for ensuring that electric lights are not needed during the day. In a cold climate a dark room will seem colder than it really is and the occupants are more likely to want to turn on heaters to compensate. Natural light, and particularly sunlight, should be provided wherever possible. If insufficient daylight is available through windows then skylights can be used to advantage, although precautions must be taken to avoid heat loss in winter and heat gain in summer.

10.5 Building materials and construction

Developing technologies are increasingly offering new materials and construction techniques which may be used in the energy efficient house. If the motivation for building such a house is the conservation of energy and not simply the provision of increased levels of thermal comfort then the implications of the use of these materials and techniques must be considered. The capital energy cost and the environmental impact of a material's production must be taken into account in the assessment of its value for energy efficiency. There is, for instance, considerable debate on the relative energy efficiencies of steel and timber in domestic construction and proponents of each side tend to adapt the statistics to support their own interests. Substantial and credible research is required to furnish designers with the information necessary to make responsible decisions in these areas (Chapter 12 addresses this in greater detail).

10.6 Building systems and services

Similar concerns may be raised with regard to the systems and services which may be installed in the energy efficient home, although in this area the choices are somewhat clearer and the ultimate energy savings more readily defined. While space heating and cooling account for a significant amount of domestic energy consumption this figure may be reduced by passive building design. Water heating in temperate climates often accounts for as much as 30% of domestic energy. There is thus a significant opportunity to reduce energy consumption simply by installing more efficient means of heating water. The greatest obstacle to this occurring (for the technology is already well-developed for solar and heat pump methods) is the high capital cost of alternatives to conventional systems and the competition from off-peak electric tariffs. Policy changes which saw the true cost of electricity charged to the consumer and incentives for use of solar systems could turn this around in a very short time. These and other aspects of appliance technology are discussed in Chapter 11.

10.7 Prototypes

The choices for effecting change in the housing patterns of the residential sector are persuasion, pricing or prescription. One of the most powerful methods of persuasion is by the production of prototypes: real buildings which designers, builders and prospective occupants can visit and where they can experience first-hand the benefits of passive design. The term prototype is used here to indicate not a design which is to be slavishly copied but one which embodies the principles of energy efficiency and is readily comprehended by the public. Such prototypes can also help dispel concerns regarding the appearance or form of the energy efficient house.

In the past many passive solar homes have been one-off designs for private individuals who may have made their homes available for public display, or to other interested parties, once or twice a year. Today many building companies have developed project homes which offer passive solar design features.

Some companies offer an energy efficient home as one of their range, thus reinforcing the idea that energy efficiency is somehow distinct from the norm and that to achieve it is either more costly or requires other qualities to be compromised. In the meantime many expensive project homes lack basic features such as roof insulation.

In many countries, such as the USA, the UK and Australia, the utilities have also seen an opportunity to build prototypes, understandably incorporating appliances which use their own form of energy rather than any more efficient alternatives. When undertaken in conjunction with a local builder this can be an effective way to market an energy efficient home.

The lack of prototypes is of particular concern when it comes to renovations. In most urban areas the number of new houses built each year is very low compared to the number of existing houses renovated, many of which are poorly designed for thermal comfort and consume large amounts of energy to compensate for their deficiencies. Unfortunately most people regard home improvements as a task within their own capabilities and so rarely seek any professional advice on their proposals. As a result many opportunities to improve the thermal performance of their dwellings are lost.

This could be changed if prototype renovations were undertaken and publicized. Real estate agents, who are largely responsible for marketing of renovated houses, could take a significant role in pointing out to purchasers the advantages of energy efficient renovations. At present their influence is underrated because they are generally uninformed in this respect.

The introduction of guidelines for energy efficiency would ensure that those responsible for new developments and renovations were aware of the opportunities to incorporate energy efficient features in their projects. Guidelines have an advantage over regulations in that they can afford to be comprehensive in scope, whereas regulations, by virtue of the fact that they must be enforceable and resistant to challenge, are relatively restricted. Guidelines may also be adapted to suit the local situation, whereas regulations are usually framed at a state or even higher level and must necessarily be generalized.

This is particularly obvious if one considers the different development densities to be found in a city. Strategies which are appropriate for low density suburbs, where solar access is relatively easily ensured, are often not applicable to high density, inner urban areas, where solar access may be restricted or even absent on certain sites. Other constraints such as heritage values or retention of streetscape character may also influence the range of solutions available in the inner urban context.

Guidelines which encourage the designer to seek opportunities for energy efficiency rather than further restricting creativity are essential. For this reason prescriptive guidelines are to be avoided.

10.8 House Energy Rating Scheme (HERS)

Perhaps one of the most important mechanisms available to encourage the implementation of energy efficiency in housing is the House Energy Rating Scheme (HERS). Such schemes have operated under various guises for a number of years, with the widest range to be found in the USA. They generally have in common the assessment of particular attributes of a home to enable the awarding of an overall rating for energy efficiency. Most schemes address new single-family construction and some also include existing housing. Only rarely are multiple dwellings considered.

Approach may be broadly categorized as one of three types: prescriptive, calculational or performance based. The first allows a home to be scored for the number of energy efficient features which it possesses and a rating to be assigned on the basis of that score. The second uses computer-based simulation to predict a building's performance in terms of energy consumption and compares this with a standard to establish a rating. The third takes the records of an existing building's energy consumption and compares this with acceptable consumption levels.

Prescriptive and calculational schemes are the most widely used. The former are often themselves developed by computer simulations which allow the establishment of relative merits of different attributes and the allocation of scores to these. Performance based schemes are only applicable to existing buildings and are therefore of limited value as a tool for forecasting performance and encouraging improvement prior to construction. They are most useful as a natural extension of an energy audit which provides the occupants of an existing home with the means to identify the areas where energy consumption could be reduced.

Ratings can either be on a pass/fail basis or be graded on a points scale (commonly 1 to 100) or star system (often 5 stars with $^1/_2$ star increments). The graded systems are preferable because they recognize effort to achieve a certain standard, even if it falls short of the ideal, and provide a basis for more subtle comparisons between different buildings.

The factors which are taken into account in the ratings vary between schemes. Building envelope is always a key consideration while some refer to thermal mass, orientation and ventilation. Space heating or cooling systems and water heating services are often addressed, but other portable appliances are usually disregarded as they may be relocated when the occupants move. Occupancy factors are rarely included other than the original point setting process.

The development of a HERS is often undertaken as a result of a local public body's desire to provide house purchasers with a means of assessing the relative energy efficiency of the houses available in the market-place. A house with a high rating has the potential to be far cheaper to run than one with a low rating. In the USA certain lenders will make increased loans on this basis, assuming that the savings on energy bills can be used to finance a greater mortgage. This works well where the climate is severe, energy consumption is high and mortgage interest rates are relatively low. In Australia, where energy bills are small compared to mortgage repayments even a large percentage saving on energy will have very little effect on ability to repay a loan.

The promotion and marketing of a HERS can be an effective way of raising public awareness of the benefits of energy efficiency. As more houses are included under the scheme greater numbers of purchasers will become aware of energy efficiency as one of the criteria for home selection. Increased awareness is likely to lead to increased demand and so the scheme becomes self-perpetuating.

This, of course, is not always the case. It is not difficult to find instances of schemes which have either failed or become dormant. Reasons for this include: lack of promotion, whether due to a shortage of funds or insufficient commitment from the sponsors; failure to obtain the active support of those involved in the design, construction, marketing and financing of houses; lack of perceived incentives to participate; inherent obstacles in the HERS, such as rigid prescriptive requirements. Successful HERS are characterized by effective marketing, industry acceptance and creative, flexible rating systems.

10.9 Regulations and standards

Support for energy conservation through government policies is an important factor in the promotion and popularization of energy efficiency. Where such support is expressed by the establishment of mandatory requirements the implementation of specific measures is assured.

The nature and extent of regulation varies between countries. Many developed countries have in place regulations relating to required insulation levels. Australia, for instance, has a number of standards related to energy efficiency but compliance is not necessarily required within the building approval process. One state has introduced mandatory R-values to be achieved by roof, walls and floors; this in effect will make insulation compulsory.

Other countries, particularly those with more severe climates and greater energy consumption, have adopted energy efficiency requirements in their building codes. The UK upgraded their Building Regulations in 1990 to include provisions for limiting heat loss and maximizing heat gain through the building fabric; controlling the output of the space and water heating systems; and limiting heat loss from water storage vessels, pipes and ducts. In the USA California has Building Energy Efficiency Standards incorporated into their building code. Compliance may be achieved by following prescriptive packages, reaching a minimum score on a points system or establishing an acceptable level of energy consumption using computer modelling of a proposed design. As in many other places with mandatory standards, California is also developing a HERS.

10.10 The role of the user

The amount of energy theoretically required to maintain a specified comfort level is one measure of the energy efficiency of a design. However, this may bear no direct relationship to actual consumption. There are many reasons for this, all generally related to the users' comfort needs and desires and their willingness to be responsible for the most efficient means of operating their home.

It is now well-documented that measures such as the installation of insulation are often used simply to raise the comfort levels of the home without any reduction in energy use. This is especially so in the temperate areas where the general thermal efficiency of houses is very low. In such cases it is difficult to adequately heat a house and so sub-standard comfort is the norm. When the thermal efficiency is improved thermal comfort becomes a realistic possibility (and at modest cost). The user often opts for increased comfort rather than energy cost savings. Relatively few people are interested in reducing energy consumption for altruistic reasons, although they may do so in the short term. Far more are attracted by the idea either of enhanced comfort or of lower energy bills.

Public education campaigns tend to focus on both these aspects in order to gain support for energy efficiency. There is often concern that people's perceptions of saving energy are bound up with notions of sacrifice, hardship and a reduced standard of living. Consequently much of the promotion of energy efficiency seeks to dispel these ideas, concentrating on the idea that energy efficiency requires no change in lifestyle. Unfortunately a side-effect of this approach is the implication that no effort is required to improve energy efficiency, thus discouraging people from taking genuine responsibility for their behaviour. It is also misleading in that a passive solar home requires some input from the occupants, even if it is only opening the curtains on winter mornings to let the sun in during the day and drawing them as soon as the sun sets to keep the heat in at night.

A better approach is to demonstrate to people the close association between amenity and energy efficiency. Most people respond positively to the idea of a house which has sunny rooms in winter and plenty of natural light, but are unaware that these qualities are inherent in the energy efficient home. It is difficult for people who have never experienced the delight of living in a house which relates well to its surroundings and responds to the climate to understand the difference that such a house could make to their quality of life. It is perhaps in this area that the need for public education is greatest.

Campaigns which are aimed solely at saving energy through conservative low- or no-cost practices have been shown to be of little value in the long term.

10.11 Design for the 21st century

There is a tendency in predicting the future to conjure extravagant scenarios very different from present experience. When we consider that most of the housing being built now will be with us well into the 21st century (and even to the end of it if it is well-maintained), then we must realize that our streetscapes will be largely unchanged for some time to come. Certainly the futuristic houses envisaged in the 1970s, with their high-tech components and self-sufficient systems, are unlikely to be realized. Change is likely to be more subtle and less visible, though perhaps none the less significant.

At the planning level we can hope to see the development of housing estates integrated with other land uses, which take advantage of the local climatic conditions to provide opportunities for energy efficient houses to be built. In the form of housing, a greater use of climatically appropriate designs and more regional variation is likely to be evident. Passive solar design can be expected to be applied more widely, with active methods limited only by research and development and cost in the market-place.

Choice of building materials may well be influenced by the outcome of research into which are the most energy efficient in production, although more complex environmental issues may need to be taken into account. The use of timber, for instance, which requires very little energy for processing compared to steel, will be constrained by availability of plantation products as it becomes increasingly unacceptable to log diminishing natural forests. More effective insulation materials will need to demonstrate a net benefit if they are high energy consumers at the stage of manufacture or if they incorporate substances which present any form of health risk. Life cycle costs of materials which are difficult to dispose of will also influence material use.

Development of sophisticated glazing for the commercial market can be expected to have a flow-on effect in the housing market as prices fall with increasing production. Already double-glazing is far more affordable at the domestic level than it has been in the past and in cold climates it could be expected to become standard. So-called 'smart' windows, which are spectrally selective, would make a significant difference to the comfort of houses in warmer climates and could mitigate the desire for air-conditioning.

In the area of services it is to be hoped that solar water heating will become more affordable and widely accepted. Provision of adequate roof or other outdoor space with optimum solar access for the collectors should now be made in all housing design. This sort of design adjustment would generally be easily made and would make little difference to the appearance of housing but could facilitate the installation of systems which would save an enormous amount of energy.

It is the actual design techniques that are likely to be subject to the greatest change as we move into the 21st century. The widespread availability of computer services is now enabling individuals to assess the thermal performance of their buildings at the design stage. Complicated calculations can be performed as part of programs which are relatively simple to run. The designer will be able to describe a given building and then vary the design characteristics of the building envelope to evaluate the effect of these changes on performance. At present the range of building types which can be modelled is somewhat limited, but can be expected to grow as more people become familiar with computer aided design and the demand for adaptable programs increases. Such changes are also likely to be spurred on by the introduction of guidelines and regulations for energy efficiency, compliance with which must be ensured during design.

Computers are an evaluative tool which will enable designers to escape the restrictions inherent in prescriptive or rule of thumb methods. For those who do not have access to computer programs other evaluative methods which are computer-generated may become available. Some researchers have already identified the need for design information to be presented in a more meaningful way than it has been in the past. This remains an on-going task for those involved in the field of energy efficient building design.

Further reading

Balcomb, J. D., Jones, R. W. McFarland, R. D. and Wray, W. O. (1984) *Passive Solar Heating Analysis, A Design Manual*, American Society of Heating, Refrigerating and Air-Conditioning Engineers, Atlanta, GA.

Butti, K. and Perlin, J. (1981) *A Golden Thread, 2500 Years of Solar Architecture and Technology*, Marion Boyars, London.

Drysdale, J. W. (1947) *Designing Houses for Australian Climates*, Experimental Building Station, Sydney, Bulletin No. 3.

Evans, M. (1979) *Housing, Climate and Comfort*, The Architectural Press, London.

Givoni, B. (1976) *Man, Climate and Architecture*, 2nd edn, Van Nostrand Reinhold, New York.

Harkness, E. L. and Mehta, M. L. (1973) *Solar Radiation Control in Buildings*, Applied Science Publishers, London.

Knowles, R. (1974) *Energy and Form: An Ecological Approach to Urban Growth*, MIT Press, Cambridge, MA.

Koenigsburger, O. H., Ingersoll, T. G., Mayhew, A. and Szokolay, S. V. (1973) *Manual of Tropical Housing and Building, Part 1: Climatic Design*, Longman, London.

Markus, T. A. and Morris, E. N. (1980) *Buildings, Climate and Energy*, Pitman, London.

Mazria, E. (1979) *The Passive Solar Energy Book*, Rodale, Emmaus, PA.

Moffar, A. S. and Schiler, M. (1981) *Landscape Design that Saves Energy*, Wm. Morrow, New York.

Olgyay, V. (1963) *Design with Climate*, Princeton University Press, Princeton, NJ.

Robinette, G. O. (ed.) (1977) *Landscape Planning for Energy Conservation*, Environmental Design Press, Reston, VA.

Rudofsky, B. (1964) *Architecture without Architects*, Museum of Modern Art, New York.

Van Straaten, J. F. (1967) *Thermal Performance of Buildings*, Elsevier, London.

Watson, D. and Labs, K. (1983) *Climatic Design*, McGraw-Hill, New York.

Technological options and sustainable energy welfare

11

Jorgen S. Norgard and Bente L. Christensen

11.1 Introduction

Today only a few million people in the world live within cultures which are environmentally sustainable. The other five billion people live in unsustainable societies which will necessarily undergo some fundamental changes in the near future, either forced upon them by acute environmental problems or – preferable – brought about through carefully considered actions. In light of this outlook the industrialized countries will have to re-examine their goals for development in technology, population, economy and attitudes in general. These elements in societies must be adapted to fit within an environmentally sustainable framework.

This chapter will briefly describe some possible ways of addressing these new challenges and will also illustrate how energy and environmental development are closely correlated with other aspects of society. Energy is an essential area for planning, as it plays a central role in the general development of technology, economy and the environment. It is, therefore, clear that energy planning for sustainability cannot be satisfactorily carried out if it is separated from the goals in economics, population, education, etc.

In section 11.2 examples are given of the environmental problems which reflect the unsustainability of the present path in the industrialized

Global Warming and the Built Environment. Edited by Robert Samuels and Deo K. Prasad. Published in 1994 by E & FN Spon, London. ISBN 0 419 19210 7.

countries. After that in section 11.3, suggestions on ways to integrate the existing environmental limitations into energy planning are provided. Finally, in the next four sections, four parameters determining the environmental impact from the use of energy in the future, namely: technology, population, the economy and individual attitudes are discussed.

11.2 Environmental sustainability

The limits to human activities in a finite world have been discussed since the 1960s and 70s (Carson, 1962; Goldsmith *et al.*, 1972; Meadows *et al.*, 1972). Since then some environmental problems appear to have been solved or postponed but new problems have emerged. Today most environmental scientists agree that present trends in development in industrialized countries cannot be sustained environmentally for very long.

11.2.1 Greenhouse effect

One of the most discussed environmental problems is the global increase in carbon dioxide in the atmosphere. Carbon dioxide, CO_2, is an inevitable by-product of burning fossil fuels, most intensively produced from burning coal and least from burning natural gas. This extra CO_2 is considered a pollutant when emitted because it increases the natural CO_2 content of the atmosphere. The enhanced CO_2 concentration will eventually cause climatic changes through increasing the greenhouse effect and raising the average global temperature.

Furthermore, CO_2 emission constitutes a pollution for which there seems to be no realistic option for cleaning at the stacks. The burning of biomass like firewood or straw also emits CO_2 but this is eventually neutralized by the equal amount of CO_2 absorbed from the atmosphere as the biomass is built up again. Only when the stock of biomass is permanently reduced, as through deforestation, is a net amount of CO_2 added to the overall atmospheric level. Table 11.1 shows estimates of the role of CO_2 and other gases contributing to the greenhouse effect.

The gases in Table 11.1 all cause higher global temperatures than would otherwise occur. The uncertainty relates only to how fast the temperature will change, whether the change has already taken effect and which climatic changes the increase in temperature will cause.

11.2.2 Other environmental impacts

The list of greenhouse gases includes CFC gases. These synthetic gases can cause an additional severe environmental hazard, namely the

Table 11.1 Estimates for the relative contribution to global warming from each of the important human-caused greenhouse gases in the 1980s (Source: UNEP, 1990). An estimated 10–30% of the CO_2 contribution comes from deforestation; the numbers are continuously updated and revised. New contributors to global warming are revealed and others, like the CFC-gases, will be phased out, but the overall picture in the table holds, with the combustion of fossil fuels as the dominant cause

Gas	Contribution (%)
CO_2 from fossil fuels and deforestation	55
Methane (CH_4)	15
CFC gases	24
N_2O	6

depletion of the ozone layer in the upper atmosphere, the stratosphere, where this layer shields life on earth against excessive ultraviolet radiation from the sun. In recent years such depletion, causing ozone 'holes', has been observed in several places around the globe.

CFC gases are used in many energy related technologies, primarily as a refrigerant in cooling systems and heat pumps, and as a propellant and insulating gas in foam materials for thermal insulation.

Energy consumption and energy technologies cause many other environmental impacts. Sulphur dioxide, SO_2, and various nitrogen oxides, NO_x, are some of the most discussed pollutants because of their damaging effect on forests, agriculture, human health and buildings. The emission of radioactive pollutants, heavy metals and certain toxic hydrocarbons are also associated with energy production. Even if these pollutants can be cleaned at the stack they still constitute a problem as a waste to be deposited elsewhere.

Nuclear power causes special environmental problems, mainly associated with a relatively low risk of accidents with extreme consequences, and with the disposal of long-lived radioactive waste. This and the high cost of nuclear power plants are the main reasons why some countries have eliminated nuclear power as an acceptable energy source.

A different kind of environmental problem is the gradual depletion of the non-renewable energy resources, moving the extraction process to more and more environmentally fragile areas, such as offshore drilling, arctic exploitation, etc.

Even the use of renewable energy resources like biomass, wind, photovoltaic, hydropower and others can cause environmental problems, although they are usually of a lower order of magnitude and of a more reversible type than those caused by non-renewable sources. Biomass sources, however, can be exploited beyond the carrying capacity of the ecosystem in question, and resulting damage to the soil can be severe and irreversible.

11.2.3 Sustainable development

The term sustainability has in recent years been used and abused a great deal by decision-makers, often in a pretentious effort to claim that they are environmentally responsible.

Economic theory provides a definition of income which includes sustainability. According to the British economist John Hicks, income is the maximum value which can be consumed during a period and still leave the consumer as well off afterwards (Daly and Cobb, 1989). If the externalities such as resource depletion and other environmental impacts are included in Hicks' definition of income, this is also a valid definition of overall sustainable consumption or, more generally, of sustainable development.

A more operational definition of an ecologically sustainable development implies that:

1. renewable resources (like biomass) should not be utilized at a rate higher than that of regeneration;
2. pollutants (like CO_2) should not be emitted at a higher rate than that which the natural environment is able to absorb and neutralize;
3. non-renewable resources (like oil, coal, gas and uranium) should not be used.

In a somewhat 'softer' definition, economists, such as Daly (1990), would add to item 3: 'at a higher rate than human-made capital can replace the natural capital lost by consuming these fuels'. This means, for instance, that if oil is used to build renewable energy systems which can provide more energy than that lost through the oil consumed, use of oil can be considered sustainable. This traditional economic view requires, however, that future values of all costs and benefits from using non-renewable resources are known, including social and environmental cost and benefits.

Around the world all three requirements for sustainable development are being violated. The high use of fossil fuels is an example of violating 2. and 3., as CO_2 is emitted at a much higher rate than the oceans and biosphere can absorb, and non-renewable resources are used at a much higher rate than capital to replace them is being built up. Generally, the activities of the global community is beyond the limits of sustainability, as is suggested in Meadows *et al.* (1992) in a 20-year follow up of *The Limits to Growth* (Meadows *et al.*, 1972).

11.2.4 UN commissions and their recommendations

In 1987 the UN commission on environment and development published the findings of a two year study in the report *Our Common Future* (World

Commission on Environment and Development, 1987). The 22 committee members from all over the world, led by the Norwegian Prime Minister Gro Harlem Brundtland, came up with some rather remarkable statements. For example, the commission strongly urged that over the next four decades primary energy consumption per capita in the industrialized countries should be reduced to about half. Furthermore, renewable energy sources should be promoted to supply an increasing part of the remaining half.

A more recent UN report (UNEP, 1990) presents evidence that the Brundtland commission report did not go far enough in its proposed measures to reach sustainable development. The report was published by the Intergovernmental Panel on Climate Change, IPCC, which was jointly established in 1988 by the World Meteorological Organization and the United Nations Environment Programme (UNEP, 1990). In October 1990 the panel concluded that in order just to stabilize the concentration of greenhouse gases the global emission level will have to be reduced quite dramatically. For CO_2 the global reduction necessary would be 60% if achieved immediately. However, it is not realistic to achieve the necessary changes in less than a few decades, so in the meantime a higher CO_2 concentration will build up, and hence the final emission reduction will be more than 60% – probably around 80% of the present level on a global basis. Similarly, the emission of other long-lived greenhouse gases will have to be reduced drastically to avoid severe climatic changes.

An exemplary international agreement has been demonstrated on phasing out CFC gases – although it might yet turn out to be too little and too late to prevent severe damage. As far as a reduction in CO_2 emission is concerned the rich countries were not, during the UNCED Earth Summit in Rio de Janeiro in 1992, able to agree on constraints in their CO_2 emissions. The European Communities have in principle agreed to stabilize their total emission by the year 2000 – a very modest goal. More international and foresighted agreements on CO_2 are still not reached.

11.2.5 Global allocation of permissible pollution

The industrialized countries are responsible for the majority of the global environmental threats being faced, such as the greenhouse effect and the ozone depletion. Due to the accumulation of emissions etc. from these countries it seems fair to allow the developing countries to emit more per capita than that of the industrialized countries for a limited period of time. A politically more realistic principle for the future could be to allocate the allowable global environmental loads on an equal per capita basis throughout the world. It can, however, be argued that the basis for this allocation should not be the future population, but rather the present population, in order to also encourage an environmentally responsible population policy.

Without going into great detail about the various pollutants the above suggestion would call for an equal allowable CO_2 emission per capita of no more than 200 kg/year from the energy sector. For a country like Denmark this would imply a 95% reduction of CO_2 emission over 40 years as suggested in a study for the Nordic countries (Brinck *et al.*, 1992).

11.3 Integrated environment and energy planning

To transform the industrialized societies from the present unsustainable state to a sustainable one is a tremendous challenge. Despite the urgency of solving some of the environmental problems this cannot realistically be achieved within a year or two. Long-term planning for the transformation should nevertheless be used as guidance for immediate actions to get on the right path towards sustainability.

Long-term planning reaches at least a couple of decades into the future. Considering the changes required it is essential to begin with an open mind. What are the ends to be achieved, what are the constraints and means available? Between the means and the ends it is important to be open minded rather than dogmatic in order to fully investigate and illuminate the realistic choices available for a democratic decision regarding which path to follow towards sustainability.

Figure 11.1 suggests a model for the process of planning for sustainable energy future.

11.3.1 Model for planning

The central element in the planning model is the creative process of building scenarios. A scenario is a picture of the future which is internally consistent in the sense that no contradictions prevent the scenario from becoming a reality. In our simple model (shown in Figure 11.1) the main negative output of the scenario is the energy consumption and its environmental impact. Three or four variable inputs are used to build a scenario for a future year to be considered. The anticipated inputs for that year are the energy intensity of the technology, the population in the region and the energy service level (or standard of living), which is determined by the economy and the general attitudes.

The planning process is an iterative one, which can be summarized as follows. A first scenario to be used as a reference for the subsequent scenarios can be built on the government's outlook for technology, population and economy, if such information is available for the future year

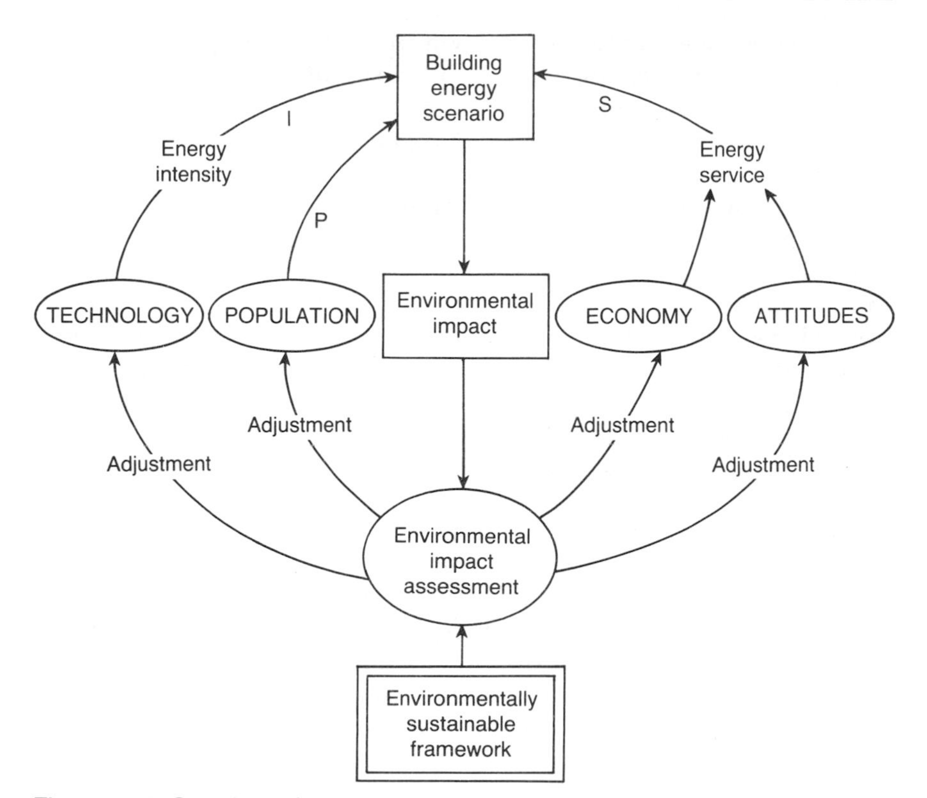

Figure 11.1 Overview of the concepts and procedures involved in an overall model for an integrated energy and environmental planning process based on iterative scenario building. The environmental impact from the resulting energy consumption is compared to the environmentally sustainable level and adjustments are made.

to be considered. The impact of people's attitudes is usually given a low priority in such a government outlook, but the importance of attitudes will be considered later in this chapter.

The level of technology in the future year should be illustrated mainly by the level of energy efficiency. In the model the efficiency is expressed by its reciprocal value, the energy intensity, I, defined as the energy required per output of energy service. In this first reference scenario the level of efficiency is based on the assumption that no special effort is directed towards improving that technical aspect.

Population, P, is essential in long-term planning for sustainability. Official prognoses or targets for population development can be used in the reference scenario.

The official economic outlooks should, if they exist for the future year considered in the reference scenario, be transferred into a set of corresponding per capita services which require energy. This energy service level is denoted by S and is described further in section 3.2.

Total energy consumption, E, can be expressed as E = PSI, which is explained in more detail elsewhere (Norgard, 1988; Norgard *et al.*, 1992). A first scenario – a reference scenario – can be established on the basis of these inputs. Its energy consumption can, with these 'business as usual' inputs, be expected to be similar to the consumption described in the official energy plan, if one exists. From the energy demand data in the scenario some concrete environmental consequences can be derived, assuming, for instance, no change in the mixture of energy sources used and no new measures taken to reduce emission at the stacks.

The next step is to compare the environmental impact of the first scenario with the critical loads of global as well as local sustainability. Most likely this first scenario will turn out to be highly unsustainable and a second scenario is built with the inputs adjusted towards sustainability. For the technology a higher energy efficiency is one of the key variables to be adjusted. A more active population policy can be considered necessary to reach a certain level of sustainable welfare. Finally, at a closer look, the anticipated future level of material welfare might be too high to be sustained. In that case the economic practice and structure, as well as the general attitudes towards material consumption, have to be adjusted towards less wasteful practice and structures in order to approach a sustainable future. The iterative process continues with new adjusted scenarios until one or – preferably – more scenarios illustrate paths to reach sustainability.

11.3.2 Energy service concept

A classic mistake in energy planning is to assume – often subconsciously – that energy consumption is a measure of welfare. This is particularly unfortunate when planning for the long term. Energy (excluding food) is of no direct value to human beings. Oil, coal, gas, uranium, petrol, electricity, etc., cannot be consumed directly. They are useful only as inputs to technologies, which then, as outputs, provide services, such as warm meals, convenient transport, comfortable rooms, illumination at the desk, clean clothes, etc. These energy services contribute directly to human welfare. The amount of energy input needed to provide the service, and hence the real value of the energy, depends on the energy efficiency of the technology involved, such as stoves, houses, trains, lamps, washing machines, etc. Figure 11.2 illustrates the energy chain of technology involved in transforming the primary energy sources into energy services. Understanding that the energy services constitute the

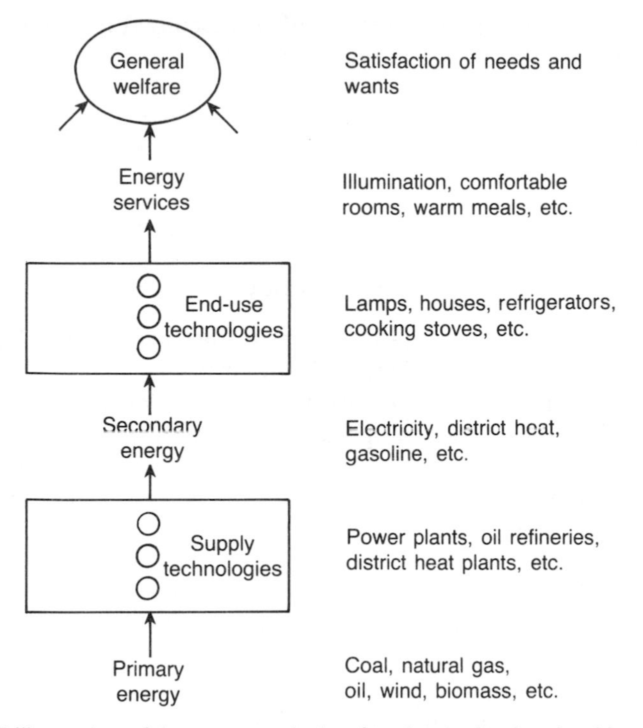

Figure 11.2 Illustration of the energy chain of technologies involved in converting primary energy into energy services and welfare. The chain can be divided into end-use and supply technologies, each of which may contain many components.

real value, rather than the energy itself, is one of the most important prerequisites in energy planning for sustainability.

The various energy services cannot usually be measured in energy units, and even then not in the same physical units. Hence, they cannot easily be added up to one single indicator of energy service. Additionally, the efficiencies of the end-use technology cannot be described by dimensionless numbers in percentages.

Planning models similar to the one described here, focusing on the energy service concept, have been used in projects for environment and energy demand in the Nordic countries (Meadows *et al.*, 1992), and for Western Europe's demand for electricity in the future (Norgard *et al.*, 1992). Despite an anticipated use of highly energy efficient technologies there are clear indications that the economic growth expected by governments, cannot, in the rich countries, be sustained environmentally in the decades to come (Benestad *et al.*, 1991; Norgard *et al.*, 1992).

Further details about the planning technique will not be described, but the content and outlook for the four determining factors in an environmentally sustainable energy future will be discussed in more detail. The four factors are:

- technology
- population
- the economy
- attitudes.

11.4 Technological options and limitations

The energy supply technologies, shown in the lower part of Figure 11.2, can be improved considerably, both in terms of making them more efficient in the process of converting primary energy into secondary energy and in terms of designing them to use renewable energy sources. Additionally, the environmental impact from the energy supply can be reduced through technological cleaning efforts. In this section the end-use technologies and the vast possibilities for improving their efficiency will be considered.

Several studies have shown that there is an enormous potential for an improvement in efficiency. In theory the potential seems unlimited in many cases. As an example, around 35% of energy consumption in industrialized countries is spent on the passive task of just maintaining temperature differences. This includes keeping houses warm inside in a cold climate, or cool inside in a warm climate, or keeping refrigerators or freezers cold in a warm environment. If thermal insulation was perfect such tasks would require no energy consumption. Similar theoretical considerations can be applied to energy used for transport and for temporary changes in temperature, such as those used in many industrial processes.

In real life, however, there are limits to the technical potential for conservation of energy. It is estimated to be realistic to expect to improve end-use efficiency on average by a factor of three over a twenty-year period. In other words, the present energy service or material standard of living can be enjoyed with just one-third of the present consumption of secondary energy, or the energy service output can be tripled without increasing the consumption of energy (Goldemberg *et al.*, 1988; Norgard and Christensen, 1982).

11.4.1 Environmental impact and cleaner technology

The major environmental impact from end-use technologies (see Figure 11.2) usually comes through their energy consumption during use. This causes emission of CO_2, SO_2, NO_x, etc., as well as waste disposal problems, resource depletion and other environmental damages and risks.

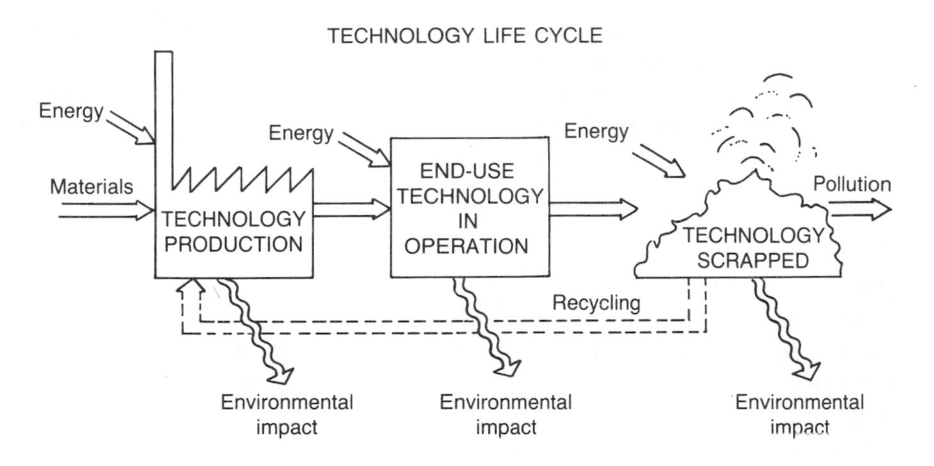

Figure 11.3 Illustrative diagram for a cleaner technology analysis. Each technological component in the energy chain shown in Figure 11.2 should, in principle, be subjected to life cycle analysis.

There are, however, other environmental problems associated with the end-use technologies. One example is the various CFC-gases presently used for insulation foam and as a refrigerant in air-conditioners, refrigerators, heat pumps and district heat pipes. These synthetic gases have a damaging effect on the ozone layer in the atmosphere, and they also contribute to the greenhouse effect.

Cleaner technology is a term used for a technology with low overall impact on the environment. An essential tool in searching for cleaner technologies is life cycle analysis. This implies studies of the environmental impact from a certain technology during its entire life cycle, including producing, running, and scrapping the technology.

A full analysis of the technology involved in providing a certain energy service can be rather complicated in principle, as indicated in Figure 11.3. A chain of energy technologies transform the primary energy into the service wanted. Each component in this chain should, in principle, be subjected to life cycle analysis, but with some experience and by establishing catalogues for energy technologies and their environmental effect the procedure can be more manageable.

The following example illustrates the importance of life cycle analyses. Rumours often arise about the production of an energy saving technology which probably consumes more energy than will be saved during its entire use! For example, this has been suggested for the compact fluorescent lamp. A life cycle analysis revealed, however, that the production energy amounts to only around 1% of what the lamp saves in its lifetime (Goldsmith *et al.*, 1972). There are, however, other environmental aspects of the compact fluorescent lamp.

General advice for designing environmentally benign appliances is to give well known materials and technologies preference over new synthetic materials and new technological systems, since the latter always can hold unforeseen, and sometimes unforeseeable, environmental risks.

11.4.2 Electricity use as an example

Electricity is used for hundreds of different tasks, each of which usually plays a relatively insignificant role in the economic budget of the home, institution, industry, or wherever electricity is used. For that reason savings options have often been ignored, even though electricity accounts for around 25% of the world's use of primary energy and the proportion is growing fast. The following examples will illustrate the vast potential for technical electricity savings. Further details and references can be found elsewhere (Norgard, 1989).

Refrigerators

A very efficient refrigerator has been developed and is now on the market, see Figure 11.4. The 200 litre unit consumes only 90 kW per year, or around 25% of what is typical for most refrigerators in use today (Benestad *et al.*, 1991; Norgard *et al.*, 1992). Still higher electric efficiencies can be achieved for refrigerators.

Whenever the purpose is to maintain temperature differences, as in the refrigerator or freezer, a high thermal insulation is always an obvious design option. Improvements in insulation can basically be achieved through thicker insulation and materials with lower thermal conductivities – or a combination of the two. The 30 mm thick polyurethane foam, which is still typical in a European refrigerator, is an inheritance from the time of almost free energy. Unfortunately it has been difficult to get manufacturers to accept thicker insulation. But the thickness should not be sacred in the search for environmental sustainable technologies, since it is a safe and reliable option. The extra space required for instance to double the insulation thickness typically amounts to less than 0.05% of the total space in a house. Insulation materials are discussed later.

Efficiency of the compressor unit, including the electric motor, is another option for improvement. The traditional reciprocal compressor and the newly introduced rotary compressor seem to be competing, and both are superior in efficiency to the absorption systems when electricity is available. Over the last two decades the efficiency of hermetically sealed reciprocal compressors has been improved by 30–50%. But still the efficiency of these small motor-compressor units is relatively low and holds the options for further improvements.

With well insulated refrigerator and freezer cabinets the problem arises

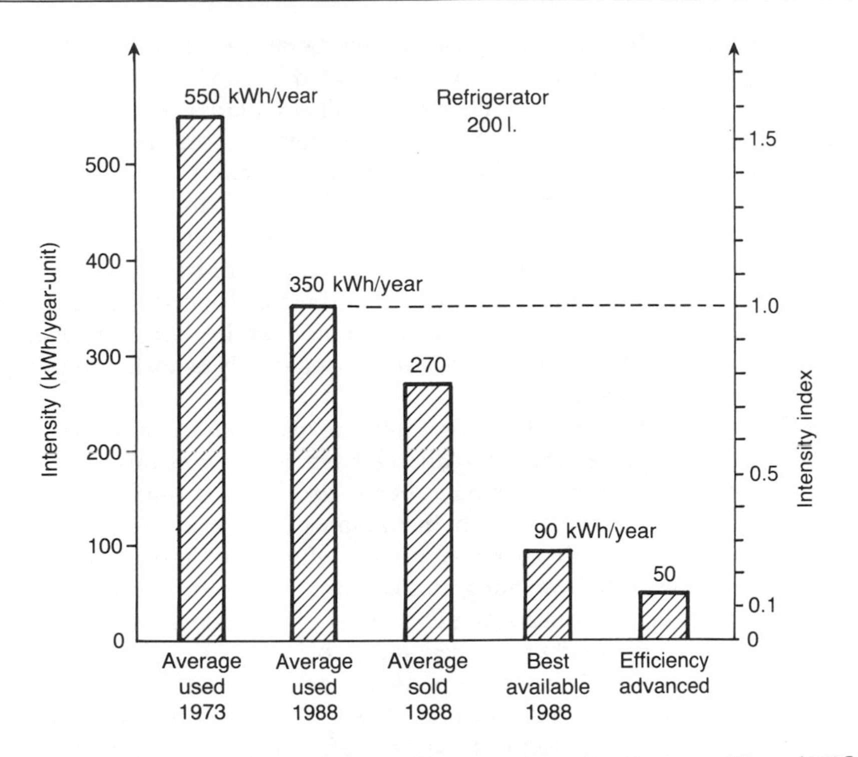

Figure 11.4 Electricity consumption at European standard test conditions (25°C outside, +5°C inside) for various versions of a 200 litre refrigerator with no freezer compartment. The intensity index on the right-hand scale shows the relative improvements, and is approximately transferable to other sizes of refrigerators, to freezers and to combined units and commercial units in shops etc. (Source: Norgard, 1989.)

that the compressors available are too large in capacity. Even the smallest unit may run only 10–15% of the time and so the two heat exchangers (evaporator and condenser) are utilized only for short periods. This causes a relative high temperature difference in the heat exchangers, which results in a low system efficiency. One solution to this problem could be to design smaller compressors, but they will usually be less efficient. Another option has been developed and successfully implemented, namely higher thermal capacities of the heat exchangers (Guldbrandsen and Norgard, 1986).

Higher thermal capacity of heat exchangers, for instance by attaching them to a steel plate, makes it possible to keep the temperature high in the evaporator and low in the condenser during the short period the compressor is running. The thermal capacities can store the heat (and cold) delivered during the short running period and dissipate it slowly during the still periods. The coefficient of performance (COP) of the system is inversely proportional to the temperature difference between the two heat exchangers. For a well insulated cabinet the COP can, in this way, be increased by typically 40%. Another advantage of the thermal heat capacity solution, rather than smaller compressors, is that a high cooling capacity is still available if occasionally needed.

If both refrigerator (+5°C) and freezer (–18°C) storage volume is requested it is often the best solution to have two separate refrigeration systems with two compressors. The average COP will be higher because of the better temperature conditions, also there will be less frost build up and the temperatures can be controlled more easily, which also saves energy. Finally, the two units can be shut off separately as well as repaired and replaced separately.

A life cycle analysis of refrigerators reveals other environmental impacts than those associated with their electricity consumption. The CFC-gases, which for decades have been used both as refrigerants and as a blowing agent in the foam insulation of the cabinets, are to be phased out due to their ozone depleting effect. But these gases also contribute to the global warming and so do the new expensive synthetic gases now being introduced as substitutes. It is interesting, therefore, to observe that all synthetic gases seem to be replaceable by simple hydrocarbons, such as butane, propane and pentane, both as refrigerants and in the foam insulation (Haaland, 1993). This could turn out to be a breakthrough for the concept of simple, safe and inexpensive solutions to environmental problems.

Clothes washing

Washing of clothes usually consists of a combination of mechanical action, chemical action and rinsing. The main advantage achieved by a washing machine is that it provides the mechanical action, and thereby exempts people from one of the most toilsome household tasks, washing clothes by hand. Often only around 20% of the electricity used by a washing machine goes into substituting the hard work – i.e. to do the mechanical action.

In many regions of the world the water for washing is heated electrically in the machine, and in that case accounts for 70–80% of the electricity consumed. The purpose of heating the water and detergent mixture (the suds) is to speed up and enforce the chemical action provided by the detergent. Finally, the rinsing serves to remove or rather dilute the dirty suds in the textiles. When it comes to hygiene a thorough rinsing is found to be more important than a high temperature.

Behaviour and 'sufficiently clean'. It is quite difficult to clearly separate the behavioural options for saving energy in washing from the technical options, partly because the task is surrounded by many habits and traditions. Rather than aiming to do laundry at certain temperatures it shall be assumed that the energy service to be provided consists of cleaned clothes.

The next question is, what is meant by cleaned clothes? How clean is sufficiently clean? Advertisements often promoted a desire for whiteness and brightness rather than cleanness. The quality of washing will not be discussed here. The potentials for lowering the electricity consumption suggested in the following do imply some changes in habits, but are not presuming a decline in the quality of the laundry.

The following data is based on the assumption that 200 laundries per year, 4 kg each, will be sufficient for a four person household. Most of what is said in the following also applies to dishwashers and clothes driers, but in general, these will not be considered as the manual work of washing dishes and hanging clothes on a line is not such a health hazard as manual clothes washing, probably the contrary.

Basic principles for efficient washing. Figure 11.5 shows electricity consumption for washing in Denmark in the years 1973 to the present, although the pattern could apply to many countries in Northern Europe. The 1988 columns refer to a typical washing pattern with equal amounts of laundry done at the temperatures 90°C, 60°C and 30°C. In these cases

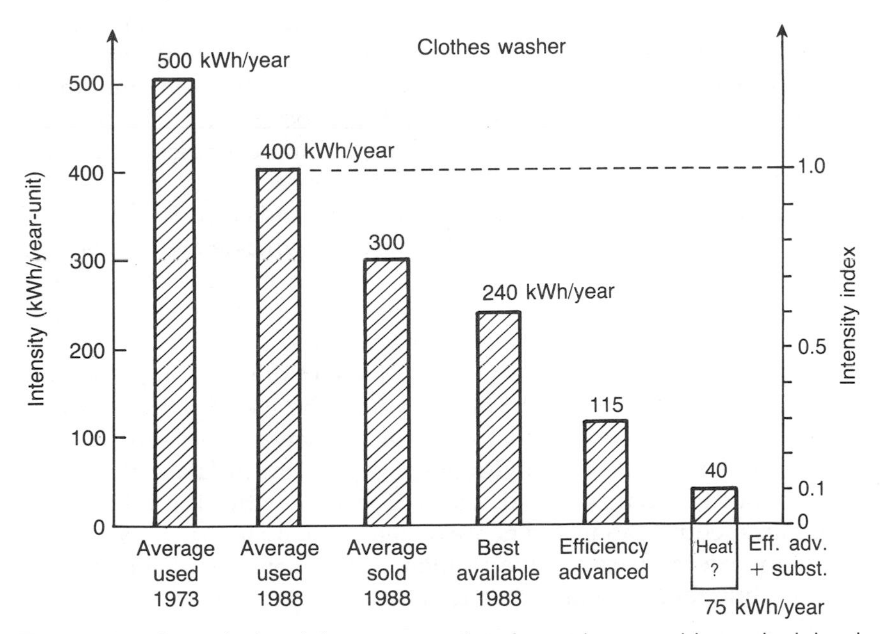

Figure 11.5 Annual electricity consumption for various washing principles in 4 kg washing machines for the 200 washings per year typical in Denmark. The differences are due to both direct technical improvements and changes in washing patterns promoted by new detergents and new fabrics (Norgard, 1989).

an obvious measure for saving electricity is to use lower temperatures, considering that typically 75% of the electricity is spent on heating the water.

The use of heat from other more appropriate sources than electricity is another sensible way to reduce the need for primary energy (as mentioned in the Introduction). This requires manually or automatically controlled intake of both hot and cold water, depending on the stage in the washing process. Warm water should be taken in only for the main wash. This feature is available and commonly used in some countries. Warm water constitutes a low quality of energy, and to use high quality energy like electricity for that purpose is wasteful from a physics point of view, and usually also in money terms.

Lower water consumption not only saves water but also detergent and energy for heating the water. Front loading machines with horizontal rotation of the drum usually consume much less water than top loaded varieties. Furthermore, proper design of the drum and the stationary vessel has already resulted in large savings of water.

Appropriate choice of washing detergent, possibly combined with an hour long soaking programme could reduce the need for high temperatures. Lukewarm or even cold water is sufficient for most laundries if a detergent with enzymes is used. Protein dissolving enzymes have been on the market for around three decades, but in recent years a fat dissolving enzyme, lipolase, has also been marketed in some countries. On the condition that it does not wrinkle the clothes a high speed spinning is an option for saving energy in the drying process which follows the washing. Finally, it is worthwhile looking for more efficient and well-controlled motor systems in all washing machines.

Electricity saving examples. The improvements indicated in Figure 11.5 (115 kWh/year) include long-term soaking in enzyme detergent, washing in low temperatures (around 30–60°C), efficient motor systems and a thorough rinsing and spinning. With warm water provided from other sources electricity consumption could be down to 40 kWh/year.

Life cycle analysis. There are important environmental impacts from washing other than the energy consumption. The use of phosphate containing washing detergents has, through the sewage systems, become one of the major causes of environmental problem in lakes, rivers and inner seas of the industrialized world. Together with nitrate the phosphate causes eutrofication and hence oxygen scarcity in the waters, which eventually exterminates the life in them. Phosphate free detergents are available but they contain other chemical compounds which may be hazardous.

In many regions of the world scarcity of water is a further concern to take into account when designing washing machines. Energy consumption and environmental impact from the production of detergents, including the new enzymes, should be watched and investigated carefully. While it seems difficult at present to suggest which detergent is the most safe to use it is certain that the use of less detergent is a positive contribution to the environment.

In general, limiting the number of washings is a safe way to reduce all the environmental problems associated with this important household activity.

11.4.3 Other uses of electricity

New light sources, e.g. compact fluorescent light bulbs, are 3–4 times as efficient as normal incandescent light bulbs. Control systems, better lamps, fixtures, etc., can also contribute in making lighting more efficient. There are examples of electricity consumption for lighting being cut to less than 20% of what is commonly in use.

Ventilation systems can also be designed to provide the same indoor air quality with only a small fraction of what present systems use, typically 15% (Norgard, 1989).

Further examples of improving electric end-use technologies can be found in Norgard (1989) and Flanigan *et al.* (1990) and the references therein.

11.4.4 Costs of saving

To save one unit of energy through more efficient end-use technology is much cheaper than producing one unit of energy. This is particularly the case when the improvements are introduced in the course of natural replacement of existing energy-consuming capital, as assumed in the following discussion. Many studies find rather low potentials for cost-effective saving through more efficient technology because the studies are often based on expensive, immediate, forced replacements of existing systems. For a long-term policy – 20–30 years – such methods for savings estimates are rather irrelevant, since most capital will already be replaced within that time horizon.

Another pitfall in cost estimates of new energy-saving technology is to assume that the price reflects the production cost. This is the case only in a perfect free market with many independent and competing manufacturers and full information being given to consumers. In the long run a higher price can reflect the extra cost, but in the first years after introducing a new energy-saving technology the price charged is often determined by the manufacturer's expectations of what kind of payback time the consumer is willing to accept.

The cost of saving energy also depends on what has already been done in this area. The more energy being saved the more expensive it becomes

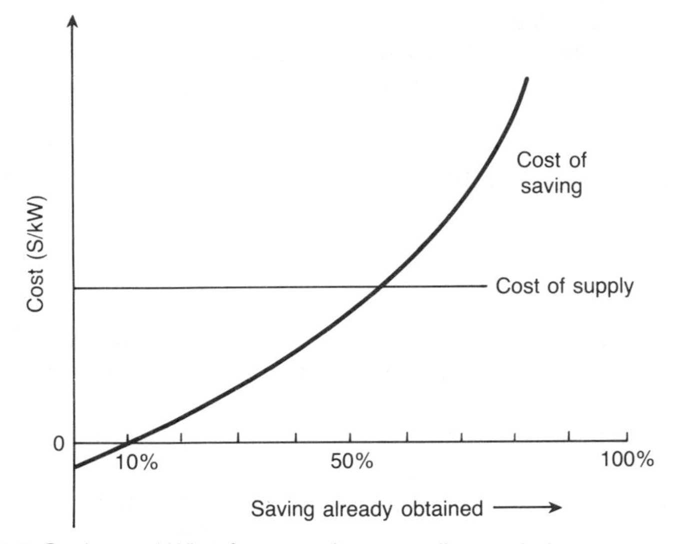

Figure 11.6 Saving a kWh of energy is generally much less expensive than supplying a kWh, shown to the left in the figure. However, the cost of saving another kWh of energy becomes more expensive as more is saved – the figure qualitatively illustrates the rising marginal cost of savings. At a certain level of saving it becomes more expensive to save than to supply energy, but typically more than half of the energy consumption in an industrialized country can be saved before that point is reached.

to save another unit, since the least expensive options are used first. Figure 11.6 illustrates this rising marginal cost of saving a kWh of electricity as compared to producing a kWh. It is estimated, as illustrated, that more than half of the present consumption of electricity can be saved with cost-effective technological measures. Only thereafter will it become more costly to save than to produce electricity. Considering the environmental benefits of saving as compared to producing there should be no hesitation in going beyond the limit of traditional economic cost-effectiveness. There should be an expectation to pay something for a better environment. In the long run, however, the environmentally benign policy of saving energy will usually also be the most cost-effective energy policy.

The fact that energy saving is cost-effective implies that the strongly needed economic development in some parts of the world, particularly in Third World countries, can be accelerated if a higher priority is given to investment in more efficient end-use technology than in energy supply technology. Frequently the type and scale of the efficient end-use technology is better suited to industrial production in developing countries than is, for instance, large-scale power plant technology.

Despite all the economic, environmental and social advantages, the energy-efficient technologies will not penetrate the market on their own, at least not at an acceptable rate. This reflects in part a lack of information on the part of decision-makers and consumers and in part an attitude favouring large-scale complex technology. It also reflects large vested interests on the part of shareholders, employees and engineers in energy supply technology. It will require a very determined sustainability policy to secure the future use of the efficient end-use technology.

11.4.5 Technology by itself is inadequate

The technical options described above for using energy more efficiently are numerous but not fully recognized. However, even these technical fixes are not adequate to secure a sustainable energy future if we otherwise continue with 'business as usual'.

One paradoxical aspect of focusing on just improving the technical efficiency is that efficiency can actually be accomplished by increasing energy consumption. For instance, if a family replaces their house with a larger house using the same standard of insulation, indoor comfort, etc., the larger house will typically use less heat per square metre of floor space, because the ratio between the heat losing surface and the service providing floor space is reduced. Therefore, the house would, according to normal technical practice, be considered more energy efficient, even though it consumes more energy. Similarly, a larger refrigerator unit will – everything else being equal – consume more electricity than a smaller one, but nevertheless also be technically more efficient, measured in electricity consumption per volume of storage space. In transportation the paradox also holds. Driving longer distances will obviously require more fuel, but usually produces better efficiency, measured as fuel consumed per km travelled.

These examples are not just accidental paradoxes. They illustrate a general view which dominates today's energy policy. Unfortunately, it is considered acceptable to increase energy consumption as long as the output – the energy service – grows ever faster, so that efficiency increases. This also applies to a nation's economy. By increasing gross domestic product, an industrialized country will usually move towards less energy intensive activities. Consequently, its energy efficiency in terms of dollars worth of output per energy unit spent will increase, although the energy consumption might increase. The absolute limits to the acceptable global energy consumption, expressed by CO_2 emissions and other environment factors, make this fixation on efficiency an obsolete view on energy conservation.

In the materialistically rich industrialized countries we must now think in terms of combining efficiency with sufficiency. An ever-growing demand for energy services will, over a period, override the environmental advantages, which efficiency improvements could otherwise have provided if the material growth had been stopped at a sensible level (Norgard, 1974). With an open mind the other variables determining the environmental impact must be discussed. What could be a reasonable population and what could be reasonable objectives for a material standard of living expressed, for example, as a per-capita energy service level?

11.5 Population – lower is better

Each country should consider its population policy in the environmental perspective of sustainability. The only serious attempt to do this, so far, seems to have been in China (Keyfitz, 1984). Even this exemplary policy of China might turn out to be insufficient in light of the global environmental problems being faced today. In any case, a *laissez-faire* population policy seems irresponsible. It is now recognized that the traditional view on population policy, i.e. that increased standard of living will automatically halt the population growth, is not going to work in many of the poorer countries. This is because of the inverse causality, namely that improving the standard of living will require a reduced population growth. Otherwise the national economic growth is more or less counterbalanced by the growth in population, resulting in no actual progress on a per capita basis, keeping the nation in a 'poverty trap'.

A population explosion in a region has historically occurred when the economic and technical development has resulted in lower death rates, but people still maintain high birth rates because of tradition, lack of information, lack of contraceptives, etc. Later a low birth rate results in stabilizing the population. Europe managed this population transition without too much peril, in part by migrating to, and conquering, other continents. It is worth keeping in mind that this option is not available for the poor, overpopulated countries of today. They are caught in a 'poverty trap' which calls for other population policy measures, like family planning programmes, social security, guaranteed old age pension, provision of other roles than childbearing for women, availability of safe and effective contraceptives and campaigning for their acceptance. Also, economic and social reward and penalty systems might be necessary and acceptable, considering the cruel alternatives of starvation, global warming, etc.

Not only the developing countries should be concerned about population problems. The industrialized countries in Europe are among the most densely populated areas in the world. This puts a lot of pressure on the natural environment. It is encouraging that population is not presently growing in this part of the world, and in fact in some industrial countries it is declining. While this is advantageous from the point of view of sustainability, the declining population is unfortunately often considered a problem by governments, mainly because of a shortsighted narrow focus on increasing the gross domestic product.

Population policy is a very emotional and sensitive area and further details will not be given here, except to emphasize the issue's extreme importance in a policy towards sustainability. The more people there are in a future world with limited sustainable energy options the lower will be the achievable average material standard of living.

11.6 Economic growth as a transition

Economic growth, measured as the annual increase in gross national product or gross domestic product, GDP, has historically been used as an indicator of the general welfare in a country. Today, increased GDP per capita is a very poor indicator of improvement in many industrialized countries. The material benefit of higher production and consumption is found to be outweighed by: the social cost of crime, drug addition, etc.; the individual cost of loneliness, alienation, distress and associated diseases, etc.; the environmental cost, threatening locally as well as globally.

Consequently, it is worthwhile considering adjusting the economic growth scenarios usually anticipated by government. In high-consuming, industrialized countries, the development should – for the sake of the humans and the environment – be focused on qualitative improvement in the economy and in everyday life, rather than on increasing material economic growth. It has already been illustrated how a lower consumption of energy, through better technology, can be achieved with a lower overall cost and hence a lower GDP and the same energy services still be provided. Furthermore, the materialistically defined energy service level is, like GDP, not necessarily a good indicator of welfare and quality of life. The end of growth in GDP is not the end of the world. An economy which at a macro level is in a steady state can at a micro level develop with just as much competition, growth, decline and other dynamic changes as do the present growth economies.

11.6.1 Temporary growth

Most economists and policy-makers today act as if economic growth must and can go on forever. They are blindly chasing the maximum, rather than searching for the optimum, size of the economy. They pursue this course because they often are forced to work within a rather short time horizon and because they do not fully perceive the long-term environmental repercussions. Their ancient idols, the pioneers in economic theory, such as John Stuart Mill, Karl Marx, Friedrich Engels,

Adam Smith and John Maynard Keynes, had a wider and longer time perspective. Hence they usually considered the economic growth as a temporary phase, a transition to a state where the economic problems of humans were solved, and all could enjoy the fruits of the technical development in a spiritually richer life (Mill, 1900; Keynes, 1931; Daly and Cobb, 1989; Norgard *et al.*, 1992). This economic state seems to have been reached for the average person in many industrialized countries, but the political determination to continue further growth still dominates economic politics.

More recently some economists, such as Daly at the World Bank (Daly and Cobb, 1989), have pointed out how unrealistic and irrational it is to pursue economic growth as a primary goal; even economic and material welfare is no longer indicated very well by the GDP. Extending the lifetime of durable goods, reuse of components and recycling of materials are examples of economic policies which might lower the traditional GDP but benefit the environment without impeding welfare, just like the already mentioned more efficient use of energy (Norgard, 1991). Another area where GDP does not reflect welfare is in leisure time. Several surveys in Europe point towards a higher preference by wage earners for more leisure time over more income, given the choice. Such an increase in welfare in the form of more leisure time would not show up as a higher GDP either. Finally, a more equal distribution of wealth and income would increase the total welfare obtained from a certain GDP. Consequently, a high degree of equity should be a basic element in a policy aimed at a sustainable economy. The declining marginal benefit and the increasing human, social and environmental cost of continuing economic growth have even turned some top politicians away from economic growth as a goal. Oscar Lafontaine, who was the Social Democratic Chancellor candidate in Germany in the 1990 election, is one of them (Lafontaine, 1985).

The above discussion illustrates how the economy, just like technology, can be made more efficient in providing welfare. This implies that better welfare could be provided with a lower GDP, constituting an important step towards environmental sustainability.

11.6.2 Poor arguments for economic growth

A valid argument for increasing production is that the products are necessary for improving the standard of living. This is the situation in developing countries, where material goods like food and decent housing are in short supply, as has earlier been the case in the presently rich countries. In the affluent societies, like Northern Europe, the situation today is quite different. In these regions it is interesting to observe that the need for more products is hardly used as an argument for economic growth. Growth is still promoted, but as a means to solve, postpone or move problems, for example, unemployment and inequities in incomes and wealth. These problems can, however, be solved in other ways if the political will is there.

Briefly, an environmentally responsible way to solve unemployment in an affluent society is to adjust working hours to the need for labour force, whether this will result in a 50-, 40-, 30- or 20-hour work week. Rather that than to increase an already environmentally straining production, and also because the latter method has, in recent decades, not been working very well. In Northern Europe the growth in industrial production does not in general create more jobs. This is explained by the fact that larger production is usually achieved through investment in labour-saving technology. Furthermore, if the increased production reflects a gain in the international or national competition with other producers then the consequence is a loss of jobs in some other place with relatively lower competitiveness.

The question of distribution is a very central issue in an economy moving towards sustainability, even though it is hardly ever mentioned. It is interesting to note that one argument for economic growth is that growth will prevent a social unrest which would otherwise be the result of inequity in a society. When the poor are better off year by year they tend to accept that inequity remains the same or even grows. This argument for growth is also mentioned in a report about the challenges in developing countries, where there are actually more valid arguments for growth (The South Commission, 1990).

A reverse argument is also put forward, namely that the inequities are necessary as a spur to economic growth. In other words, growth and inequities support each other and when sustainability calls for a slow down and eventually a halt in economic growth, then the growth argument for inequity obviously disappears.

With the recognition of environmentally limited potential for production in the world the question of distribution becomes more acute, whether referring to distribution of work, income or wealth. A democratic political answer is likely to point towards more material equity than today. The argument for growth as a means to make inequities acceptable originated in a time when the natural basis for economic activity was perceived as unlimited. The argument is irresponsible in an economy aiming for sustainable welfare.

11.7 Attitudes for sustainability

As early as 1972 the executive committee of the Club of Rome stated the wish '... that man must explore himself – his goals and values – as much as the work he seeks to change' (Meadows *et al.*, 1972). The environmentally determined need to recognize and change some of the basic social values and attitudes in the industrialized countries is even more acute today.

11.7.1 Values and attitudes

Value means here the often subconsciously held fundamental beliefs on which one's way of life is based. Examples of values to illustrate the concept are: freedom, collectivism, individualism, orderliness, traditionalism and equality. They represent emotions and are often determined in early childhood and held for a lifetime. An attitude is usually a more conscious concept and directed towards a specific situation.

Values and attitudes are not changed overnight. They are shaped during the entire process of socialization (Figure 11.7). Upbringing in preschool age often creates the emotional value pattern which constitutes the base for the rest of one's life. More concrete values and attitudes are imposed during the formal education period in schools and other institutions, as well as through TV and other media (Christensen and Norgard, 1976).

People's attitudes and social values determine how they behave within the physical frame provided by nature, the economy, technology, etc. However, the human-made structures are established and developed according to the attitudes of the people in control of designing them. In an ideal democracy these controls are held by the general public and their elected representatives. But in the real, complicated world of today not so much the politicians, but the experts in technology, economics, etc., often hold the key to decisions, and their attitudes towards the

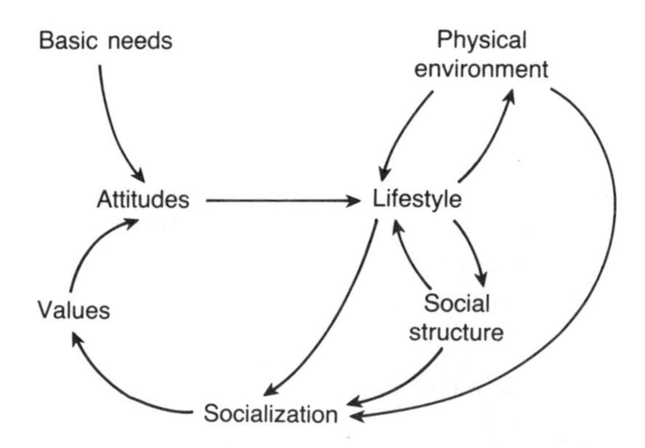

Figure 11.7 Values and attitudes are shaped by the social structure as well as the other way round. (Source: Christensen and Norgard, 1976.)

environment, technology, distribution, and so on can, and do, play a dominant role.

11.7.2 Present trends in West European attitudes

There are many indications that people in the affluent, industrialized countries are slowly changing their attitudes away from what has been the foundation in generating the high level of material wealth. Materialism is gradually being replaced by post-materialistic values and attitudes. Typical trends are that work and the economic standard of living play a smaller role than 30 years ago. These are also some of the trends in attitudes required to approach sustainable welfare.

One of the better illustrations of trends in attitudes is provided by a systematic series of opinion polls of 2000–4000 people in Denmark by the Danish National Institute of Social Research in 1964, 1975 and 1987 (Kuhl and Koch-Nielsen, 1976; Kuhl, 1980; Platz, 1988). These surveys make it possible to look at the development of attitudes over time. The question asked was, 'If you had the choice between one extra hour of leisure per day (same income) and one extra hour's pay (same working time) which one would you prefer?'. The results are shown by the three shaded columns in Figure 11.8; there is a clear increase in preference for

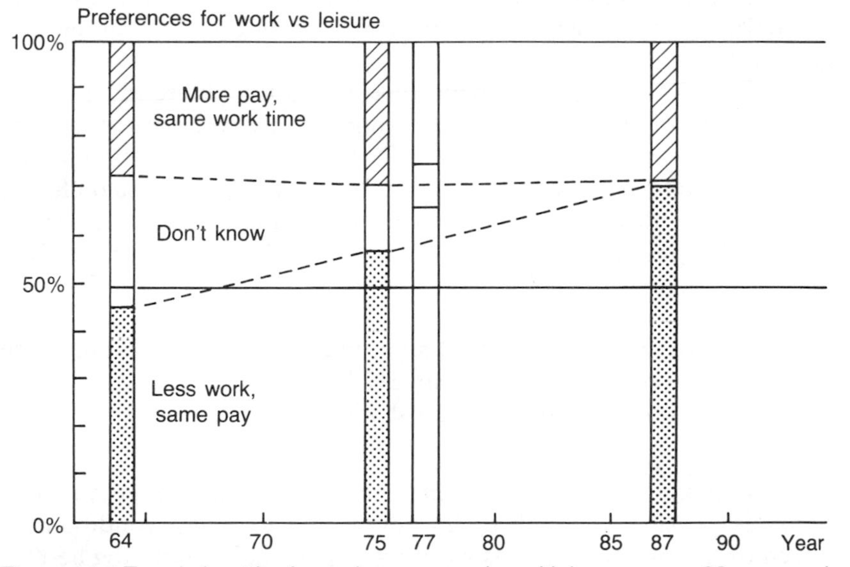

Figure 11.8 Trends in attitudes to income, work and leisure over a 23-year period of people in Denmark. The shaded columns represent the studies carried out by the Danish National Institute of Social Research (Kuhl and Koch-Nielsen, 1976; Kuhl, 1980; Platz, 1988) and the unshaded one a study by the EEC (Clutterbuck, 1979).

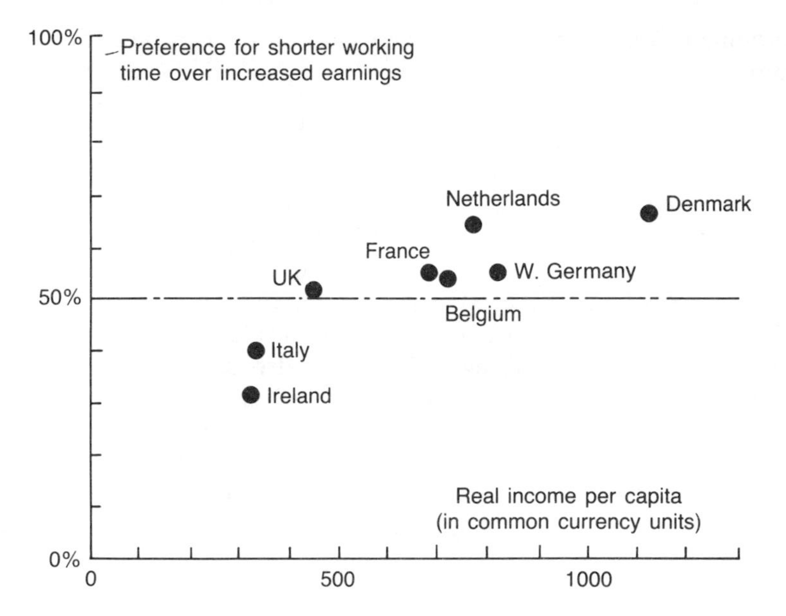

Figure 11.9 Percentage of wage earners who prefer shorter working time over more pay as a function of income and country in 1977 (Clutterbuck, 1979).

more leisure over more pay. By 1987 the ratio between the two had reached 70/29. The unshaded column represents an EEC survey (see Figure 11.9).

Other surveys in Denmark have posed slightly differently phrased questions, and the results are therefore not directly comparable, but the trends are the same.

Similar trends have also been noted in Norway. In 1982 58% preferred shorter working days over higher income according to a relatively small sample by telephone survey. Another survey in 1975 indicated that 76% found the standard of living in Norway too high (Norgard and Christensen, 1984).

Figure 11.9 shows some results of an opinion survey carried out in the European Economic Community in 1977 (Clutterbuck, 1979; OECD, 1982). The question posed was almost identical to that in the Danish survey:

> Supposing the economic situation changed for the better and it becomes possible to think of improvements in living conditions, which of the two possibilities would you personally think the best?
>
> 1. Increase in pay (for the same hours of work)?
> 2. Shorter working time (for the same pay)?

The results shown are for what is termed the 'working population', but the survey showed no significant differences when the rest of the population was included. For all countries taken together 51% of the working population surveyed preferred shorter working time, while 45% chose more pay. The results in Figure 11.8 indicate a correlation between the preference for more leisure time than an increased level of income.

In most countries the majority favoured increased leisure. For Denmark this option was chosen by 66% of the working population surveyed, while only 26% preferred more pay, which is in good agreement with the Danish survey. These surveys (along with other indicators not discussed here) imply that attitudes in some of the high-income countries are already prepared for a saturation economy, which is a prerequisite for a sustainable future.

11.7.3 Saturation promoting factors

The level of the living standard in Scandinavia, one of the highest in the world, is obviously one important factor. The higher the standard of living obtained the less there can be gained from further increases, and eventually the marginal cost in the form of work time etc. comes to be perceived as higher than the marginal benefit.

Distribution of income and wealth is fairly even in Scandinavia, which could contribute to the presence of these attitudes – although there are significant differences in incomes and wealth the inequalities are considerably smaller than in most industrialized countries.

For decades the Scandinavian countries have provided a high level of social welfare in the form of essentially free education and health care, unemployment pay and a guaranteed retirement pay for everyone. For most Scandinavians this has produced so deep a feeling of economic security that they have been able to take these benefits for granted. If they were to have to provide for themselves and their families in such a way as to guarantee this range of socio-economic security there would be almost no limit to how much wealth they would want to accumulate.

The high standard of social welfare is of course partly responsible for the high tax level in Scandinavia. Average marginal income tax rate in Denmark is more than 50%, which means that for the individual the marginal disposable income from working is rather small. For example, this high marginal income tax rate, compounded with various sales taxes, means that the average Danish car owner must work between one and two weeks per month in order to pay for and operate a car (Norgard and Christensen, 1982). Such effects no doubt help to explain why so many people prefer more leisure over more income.

Finally, the roles of women in society can explain some of the trends. In the surveys for Denmark made in 1975 (Figure 11.8) in which 57% of all wage earners preferred less work, the number of female wage earners preferring less work was 62% and males 54%. This would seem to support the general view that the values supporting a saturation society are those traditionally considered female. And, in fact, women do seem to be playing a larger role in shaping society in Scandinavia than in most other industrialized countries. In the 1980s women made up more than 40% of the labour force in Denmark.

Trends like those described above are no longer encouraged by the government in any country, because the politicians are still promoting economic growth. During the 1980s, for instance, a conservative Danish government was very slowly and cautiously reversing most of the factors which promoted these attitudes for sustainability. Social security was being lowered, income equity reduced, marginal taxes reduced and advertisements to encourage more consumption were increased dramatically and were also allowed on television.

While economic growth is needed in the developing countries a saturation in economic and material development should be encouraged in the industrialized countries, particularly in those with the highest economic standard of living. Examples of values to be promoted or encouraged for the upcoming generation are equity, sharing and leisure, while competition, material comfort and work should be somewhat de-emphasized. But it is all a question of balance, not shifting from one extreme to the other, as described elsewhere (Christensen and Norgard, 1976).

11.7.4 Attitudes directly related to energy policy

As mentioned earlier, the attitudes of the people directly in charge of energy policy will have a very profound and immediate impact on the choices in that area (Norgard and Christensen, 1988). There is no doubt that energy conservation through more efficient technology from a rational point of view is preferable to a greater energy supply. Nevertheless the policy is not being pursued very vigorously in any country. Some of the most severe barriers for this first step in a sustainable energy policy are psychological in nature, reflecting the attitudes of the decision-makers.

For engineers there has, from a traditional view-point, been less glamour and less prestige associated with electricity saving technology compared to the building of large power plants, hydro-electric schemes or other spectacular electricity supply systems. The engineering skills required for developing electricity saving systems have a more holistic and interdisciplinary character than the traditional engineering expertise. These skills have only recently emerged and begun to be recognized and evaluated. The electricity saving technologies can be equally, or more, sophisticated in their modest, interdisciplinary and integrated way, than dominant, large-scale supply technology.

Also, with their natural science background, engineers generally prefer exact definitions and achievements which can be measured. Total efficiency in the energy chain, from primary energy to human welfare, as shown in Figure 11.2, cannot be defined precisely in numbers. The efficiency of converting coal into electricity can be measured precisely, even in a dimensionless number.

Efficiency in the higher steps, the end-use technology, can possibly be expressed quantitatively, but in various units, such as in litres of gasoline per 100 km for a car and in m^2 of heated floor space in housing. But the units are different and the data still cannot be added into one single number of efficiency for a household, an industry or a nation. Such difficulties in measuring the achievements in precise numbers might make the challenges less attractive to traditional technologists, leaving behind huge potentials for saving energy.

Economists and politicians also prefer to deal with aspects which can be easily expressed in numbers. Furthermore, the monumental aspect of large scale energy technology, like a big power plant, attracts the politicians as symbols of achievements. Energy conservation and sustainability is not nearly as spectacular in the news media. This might subconsciously give these goals a lower priority in the mind of a traditional politician.

New attitudes are needed both for those directly in charge and the general public if a sustainable welfare is to be reached. Such changes are usually very slow, and might constitute the most severe barrier for solving acute environmental problems in time.

11.8 Summary and conclusions

Integrated energy and environmental planning will no doubt be an essential element in future policies as it becomes clear that the constraints on development will increasingly be the natural resources and the environment. The present way of life in the industrialized part of the world cannot be sustained for many more decades and dramatic changes will be needed. At the same time, it is crucial that the rest of the world does not follow the same unsustainable pattern of life.

In order to obtain a high welfare within the constraints of sustainability it is necessary to turn around the energy planning process, so that focus will be on end-use demand and efficiency rather than on a greater supply of energy. The technical options for reducing energy demand are enormous and must be exploited, but with a long-term global view it is at the same time clear that technology alone will be inadequate to bring us towards sustainability. Other factors in society which influence energy demand and cause the environmental problems will also have to be adjusted. These other factors can briefly be described as growth in population and growth in material standard of living, or in GDP.

Population is growing mainly in developing countries, but it is recognized there as a problem and actions are being taken to solve it. Economy is in absolute figures growing mainly in the industrialized countries, but creating environmental problems worldwide. Only recently has economic growth become recognized as a problem by a few economists and politicians. The attitudes of the public in most of the well-off countries do not appear to be a serious barrier for reaching a steady-state economy in those countries.

An encouraging official step towards environmental sustainability was taken in 1990 by the Danish government in establishing an energy plan with focus on the demand-side (Danish Ministry of Energy, 1990). The result is, in one of the three scenarios – a 40% reduction in primary energy consumption by the year 2030 as compared to now. Over the same period the CO_2 emission could be reduced by 65% without using nuclear power or hydropower. Nevertheless, as outlined in section 11.2, this is not sufficient to reach sustainability. In addition, by the year 2030 the primary energy consumption in this plan is – after the decline – growing again due to the continuous economic growth.

Referring to Figure 11.1, the Danish government scenario can be described as follows. The technological options for reaching a low energy intensity is utilized to a rather high degree, which accounts for most of the decline achieved in energy consumption and CO_2 emissions. Population is assumed to continue at present trends, resulting in 8% fewer Danes by 2030, which is another positive contributing factor, although not promoted. The economy is anticipated to continue to grow, but at a modest growth rate, 2% per year in the near future and later 1.6% per year. By 2030, however, this amounts to a doubling in per capita production and consumption. This trend is in conflict with some half-hearted efforts by government to change people's attitudes towards being more environmentally conscious. The economic growth and the corresponding growth in energy service is what causes the growth in energy demand at the end of the planning period and what in general impedes the sustainability of that government plan. However, it is encouraging that such a plan, with a drastic decline in energy consumption, can achieve broad political support. But it is discouraging and frightening that most politicians still refuse to even talk about the last factor for reaching sustainability, bringing the macro economy to a halt at an optimum level instead of striving indefinitely for a higher level.

References

Benestad, O., Kristiansen, A., Slevig, E. *et al.* (1991) Energy 2030 – Low Energy Scenarios for Denmark, Norway and Sweden. Alternative Framtid, Sognveien 70, N–0855, Oslo, Norway (in Norwegian).

Brinck, L., Emborg, L., Juul-Kristensen, B. *et al.* (1992) Environment and Energy in Scandinavia – Energy Scenarios for the Year 2010. Nordiske Seminar – og Arbejdsrapporter No. 1992: 548. Nordisk Ministerrad, Copenhagen, Denmark (in Scandinavian).

Carson, R. (1962) *Silent Spring*, Houghton Mifflin, Boston, MA.

Christensen, B. L. and Norgard, J. S. (1976) Social values and the limits to growth. *Technological Forecasting and Social Change*, **9**, 411.

Clutterbuck, D. (1979) Shorter working: more jobs or more problems? *International Management*, **34** (5), 23–6.

Daly, H. (1990) Towards some operational principles of sustainable development. *Ecological Economics*, **2**, 1–6.

Daly, H. and Cobb, J. B. (1989) *For the Common Good*, Beacon Press, Boston, MA.

Danish Ministry of Energy (1990) Energy 2000 – A Plan of Action for Sustainable Development. Danish Energy Agency, Copenhagen.

Flanigan, T., Lovins, A. *et al.* (1990) *Competitek – Advanced Techniques for Electric Efficiency*, Information Service, Rocky Mountain Institute, CO.

Goldemberg, J., Johansson, T., Reddy, A. and Williams, R. (1988) *Energy for a Sustainable World*, Wiley Eastern Ltd, New Delhi.

Goldsmith, E., Allen, R., Allaby, M. *et al.* (1972) A blueprint of survival. *The Ecologist* (special issue).

Guldbrandsen, T. and Norgard, J. S. (1986) Achieving substantially reduced energy consumption in European type refrigerators. Presented at the 37th Annual International Appliance Technical Conference, May 1986, Purdue University, West Lafayette, Indiana, USA.

Gydesen, A., Maimann, D., Pedersen, P. B. *et al.* (1990) Cleaner Technology in the Energy Field. Report No. 138. Danish Ministry of Environment, Copenhagen (in Danish).

Haaland, T. (1993) *Various Materials on Hydrocarbons in Refrigerators*, Greenpeace Denmark, Copenhagen.

Keynes, J. M. (1931) *Essays in Persuasion*, Macmillan, London.

Keyfitz, N. (1984) The population of China. *Scientific American*, **250** (2), 22–31.

Kuhl, P.-H. (1980) *Fritid 1964–75*, The Danish National Institute of Social Research, Copenhagen (in Danish).

Kuhl, P. -H. and Koch-Nielsen, I. (1976) *Fritid 1975*, The Danish National Institute of Social Research, Copenhagen (in Danish).

Lafontaine, O. (1985) *Der andere Fortschrift: Verantwortung statt Verweigerung*, Hoffman and Campe Verlag, Hamburg (in German).

Meadows, D. H., Meadows, D. L., Randers, J. and Behrens, W. W. (1972) *The Limits to Growth*, Universe Books, New York.

Meadows, D. H., Meadows, D. L. and Randers, J. (1992) *Beyond the Limits – Global Collapse or Sustainable Future?*, Earthscan Publishers Ltd, London.

Mill, J. S. (1900) *Principles of Political Economy*, Revised edn, Vol. 2, Colonial Press, New York.

Norgard, J. S. (1974) Technological and Social Measures to Conserve Energy. Research Program on Technology and Public Policy, Dartmouth College, USA.

Norgard, J. S. (1988) *Methods and Concepts in Electricity Conservation Planning*, Technical University of Denmark, Lyngby.

Norgard, J. S. (1989) Low electricity appliances – options for the future, in *Electricity* (eds T. B. Johansson, B. Bodlund and R. H. Williams), Lund University Press, Sweden.

Norgard, J. S. (1991) National product or welfare?, in Det Rene Svineri. EVA Report. Samfundsfagsnyt, Copenhagen (in Danish).

Norgard, J. S. and Christensen, B. L. (1982) *Energihusholdning, Husholdning, Holdning*, Taastrup, Copenhagen (in Danish). (Available in English as *Managing Energy – Managing the Home*, Technical University of Denmark, Lyngby.)

Norgard, J. S. and Christensen, B. L. (1984) Individual attitudes in Scandinavia point towards a low-energy saturated society, in Proceedings from the ACEEE Conference Study of Energy Efficiency in Buildings, Vol. F, August, Santa Cruz, CA.

Norgard, J. S. and Christensen, B. L. (1988) Sociological and psychological barriers to electricity savings, in *Demand-side Management and Electricity End-use Efficiency*, Kluwer Academic Publishers, Dordrecht.

Norgard, J. S., Viegand, J. *et al.* (1992) Low Electricity Europe – Sustainable Options. Balaton Report. Technical University of Denmark, Lyngby.

OECD (1982) *Labor Supply, Growth Constraints and Work Sharing*, OECD Publication Office, Paris.

Platz, M. (1988) Arbejdstid 1987, The Danish National Institute of Social Research, Copenhagen (in Danish).

The South Commission (1990) *The Challenge to the South*, Oxford University Press, Oxford.

UNEP (1990) Scientific Assessment of Climate Change. Policy-maker's summary of the Report of Working Group I, Intergovernmental Panel on Climate Change (IPCC), WMO and UNEP.

World Commission on Environment and Development (1987) *Our Common Future*, Oxford University Press, Oxford.

Materials selection and energy efficiency 12

George Baird

12.1 Introduction

Over the last decade or so energy efficiency has become of increasing concern to those who design, operate and use buildings. That this concern has been focused almost exclusively on the energy required to operate the building is not surprising. Operating energy costs are readily apparent and easily understood – the kilowatt hour and the dollar are currencies familiar to all.

Much less attention has been given to the energy embodied in the materials of which buildings are composed and the implications of their selection for energy efficiency. Virtually ignored until recently, because their effects tend to be indirect and in many cases remote from the building site, are the environmental impacts of the selection of different building materials.

The main aim of this chapter is to describe the underlying concepts and practical methods for estimating the energy embodied in a building – its capital energy cost. The text will also summarize the status of work in this field and touch on related environmental impact issues.

The terms capital energy cost and capital energy requirement are used interchangeably in this chapter, normally when referring to the total energy embodied in a building.

Global Warming and the Built Environment. Edited by Robert Samuels and Deo K. Prasad. Published in 1994 by E & F N Spon, London. ISBN 0 419 19210 7.

12.1.1 Concern with energy capital costs

While the topic of the capital energy costs of buildings has been raised from time to time over a number of years comparatively little work of real substance has been carried out. The modest literature in the field all tends to refer back to relatively few substantial pieces of research. Many people have seen it as a worthwhile area of investigation but few, having seen the enormity of the task, have been able to pursue it in any depth.

During the 1970s a general concern for the global environment prompted some observers to comment on this issue, with reference both to the energy requirements of different building materials and the environmental impacts resulting from their production and use (see Mackillop, 1972, for example). Researchers such as Chapman (1975) were calling for fuel and material conservation to become a major part of good design. However, these concerns were largely ignored by architects even though most would have regarded their profession as dedicated to the improvement of the environment rather than its degradation. The Vales (Vale and Vale, 1991) see this failure to link architecture with a concern for resources as going back at least to the origins of the classical traditions of Greek architecture.

In terms of buildings, the main research and development effort worldwide has been concerned with their design and operation to reduce operating energy consumption. A limited number of studies of capital energy consumption have also been undertaken (Haseltine, 1975; Kegel, 1975; Pick and Becker, 1975; Stein *et al.*, 1976, 1977, 1980; Stein, 1977a, b; Baird and Chan, 1983) but the implications of these were largely ignored by the mainstream of building producers.

It has to be said that conflicting opinions were evident even amongst the researchers. On the one hand, Ballantyne (1976), on the basis of his estimates that the operating energy costs of a dwelling would be about 20 times the capital energy costs, suggested that concentrating effort on the former would be more rewarding from the point of view of reducing overall energy consumption. Wright and Gardiner (1979), on the other hand, disputed the suggestion that the capital energy requirement of housing was comparatively unimportant, arguing first that, 'in absolute terms, all means of energy conservation must be sought' and second that 'whilst it is true that for some designs the capital energy requirement only amounts to a few years of operation, in other cases (they) are of comparable magnitude', ranging from 8 to 70 years.

Whatever the merits of the respective arguments it is evident that as improvements are made to the operating efficiency of buildings their capital energy requirements will become comparatively more significant. Kohler (1987), in a study of 30 buildings, estimated that the capital energy costs (including repairs) could amount to 30–40% of the operating

Table 12.1 Global sources of carbon dioxide emissions (Source: Ministry of Commerce, 1990)

Source	Contribution (%)
Oil	29
Coal	29
Natural gas	11
Deforestation	20
Other	10

energy costs of a 'low energy building' over an 80-year life. For some building types it has been suggested (Connaughton, 1992) that capital and operating energy costs could be on a par with one another over a building's life cycle as a result of refurbishments, which can be as frequent as every 10–15 years in the case of offices and shops.

12.1.2 Greenhouse gases and energy efficiency

In more recent times environmental concerns have focused on the implications of rising concentrations of the greenhouse gases in the earth's atmosphere as a result of the increased consumption of fossil fuels. This is a trend in which the building construction industry plays a significant role.

From a global perspective the major contribution to these emissions comes from the burning of fossil fuels, with oil and coal taking the main proportion, as indicated in Table 12.1.

The majority of conventional building materials are fossil-fuel intensive in their production and thus contribute directly to the rising greenhouse gas concentrations; for example, cement, fired brick, glass, steel, aluminium and many plastics. Timber, while it can be subject to energy intensive processes, also has potentially beneficial implications for the concentration of greenhouse gases in the earth's atmosphere in terms of its place in the carbon cycle.

A few attempts have been made to quantify the emission of carbon dioxide and other pollutants resulting from the burning of fossil fuels in the production of building materials and assemblies. Kohler (1987) has tabulated the rate of production of a range of pollutants for over 60 building materials and processes and has estimated the average pollution (both direct and indirect) over the life of 30 buildings in Switzerland. Cole and Rousseau (1992) have conducted an air pollution audit of four Canadian wall assemblies in terms of emissions of carbon dioxide, particulate matter (energy and non-energy related) and hydrocarbons. In New Zealand, Honey and Buchanan (1992) have made estimates of the carbon coefficients of a range of building materials and thence the carbon emissions for a small number of building types.

What is becoming increasingly clear, as the articulation of concerns about the global environment gathers momentum, is that the irresponsible practices of the past cannot continue. The effects of some of these practices may not be fully reversible and there is a strengthening call for changes, the ultimate aim of which would be to achieve a sustainable future on our planet. Unfortunately, for those who would wish to take action on these issues, there is very little reliable information on the energy intensity and environmental impact of building materials.

While climatic change seems inevitable as a result of the emission of greenhouse gases its rate and extent cannot be estimated with any certainty – the reaction of the biosphere is impossible to predict. However, there is no doubt that the burning of fossil fuels is the main source of the greenhouse gases and that it would be most desirable to use less in the construction and operation of buildings.

As far as the special case of the use of wood as a building material is concerned the situation is more complex. Despite its image as a 'green' material the clearing of forests releases large quantities of carbon into the atmosphere; only during its growing phase will a forest be a net absorber of carbon.

Not only does carbon dioxide have the highest share (at around 70%) in the greenhouse gas effect attributable to human activities, but it has also contributed the greatest proportion (45–50%) of the increase since pre-industrial times (Ministry for the Environment, 1990). The Intergovernmental Panel on Climatic Change (IPCC) has suggested that a 60–80% reduction in carbon dioxide emission would be required to stabilize concentrations at present day levels.

Despite the uncertainties inherent in attempting to predict the precise impact of these emissions on the climate of the globe it will be evident, even to the most sceptical of observers, that reductions in the direct use of coal, gas and petroleum products are imperative. To do so requires strategies dealing not only with energy production but also with its end-use in all spheres of human activity. Our concern in this book is to highlight the strategies relevant to the built environment.

12.1.3 Building energy efficiency and energy capital costs

Generally speaking, buildings have a comparatively long life. During that time, considerable quantities of energy will be used in them, the amount depending on such factors as the hours of operation, the climate, the services provided and so on. How efficiently the energy is used is to a great extent dependent on building management, but basic to the issue is the inherent efficiency of the building design itself.

The factors that contribute to energy efficient building design and energy efficient building operation are dealt with in several other parts of this book. In this chapter the concern for energy efficiency goes even deeper and will probe into the capital energy requirements of buildings. The main effort will concentrate on the energy implications of using different building materials, with comment on the general environmental consequences where appropriate.

The topic will be introduced by outlining the concept of capital energy and the terminology, conventions and limitations of energy analysis. Following that, to give an indication of national scale, the energy consumed by the construction industries of two countries, the United States of America and New Zealand, is examined and compared.

The processes involved, their likely environmental impact and the energy consumed in the manufacture of some common building materials are then described and the energy coefficients of these building materials presented. The application of these energy coefficients to the estimation of the capital energy requirements of a standard house is then illustrated and the effects of using different materials and components explored; differences between countries are also noted.

The chapter concludes with some thoughts on how to take account of these matters now and what information is needed to improve procedures for the future.

12.2 Capital energy concepts

The capital energy cost or requirement of a product is the sum total of all the energy embodied throughout the various stages of its manufacture, from raw material to finished product.

To enable the consistent estimation of these figures formalized methods (termed energy analysis) have been developed for calculating the overall amount of energy required to produce goods and services. These energy requirements include both the energy consumed directly by the particular process in question (the direct energy requirement) and the energy used indirectly to extract the raw materials associated with the process (the indirect energy).

In the case of a building, for example, the direct energy is the fossil fuel and electricity used at the building site and in transporting the materials to the site. The indirect energy is that used to manufacture materials, components, tools and machinery, plus the energy required to extract the raw materials plus the energy used to manufacture the equipment used to extract the raw materials and so on.

12.2.1 Terms and definitions of energy analysis

The most widely accepted methods of estimating a product's energy requirements are those established by the International Federation of

Table 12.2 Key terms, definitions and system boundaries in energy analysis

IFIAS level	Gross energy requirement (%)	System boundaries and definitions	Terminology used	
1	<50%	The direct energy required to carry out a given process – also known as the process energy requirement (PER)	Direct energy requirement	Gross energy requirement (GER)
2	*c.* 40%	Energy involved in extracting materials	Indirect energy requirement	
3	Rarely >10%	Energy needed to make capital equipment		
4	Very low	Energy to make the machines that make the capital equipment		

The GER percentages are a guide only, IFIAS (1975); the term energy coefficient is defined as the energy required to produce a unit of material.

Institutes for Advanced Study, and these are adopted here. The various IFIAS Workshops (IFIAS, 1974, 1975) are the source of the definitions and of the suggested levels and system boundaries to be used in respect of the gross energy requirement of a product (see Table 12.2).

It is important to note that the methods of energy analysis are still in the process of development and that those recommended by IFIAS would not normally include the energy required for life support (i.e. human energy) or take account of any environmental impact resulting from, say, the use of particular materials. In this chapter consideration of the latter will be integrated with discussion of the energy issues where appropriate (see section 12.4 in particular).

There are several methods of energy analysis available. The two most common methods were used to obtain the data presented in this chapter, namely, input–output analysis, which involves estimating the ratio of the energy input to the productive output of each industry, and pro-

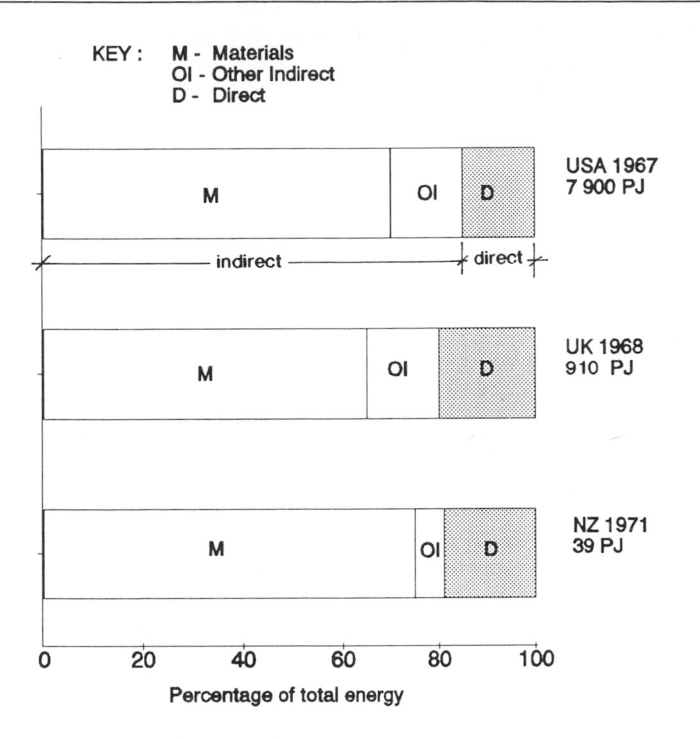

Figure 12.1 Energy requirements of the USA, UK and NZ construction industries.

cess analysis in which the energy input at each stage of a process is summed.

12.3 Energy use in the construction industry

In most industrialized countries the construction industry is largely factory based, with many processes carried out remote from the building site, work there consisting mainly of component assembly. This situation is reflected in the low percentages of direct energy required by the construction industries of the USA, the UK and New Zealand, for example, 20% or less in all three cases, as shown in Figure 12.1.

Table 12.3 gives a breakdown of the energy used to produce various types of new buildings in the USA and New Zealand during comparable periods. In both cases almost half the energy requirements for new buildings went to residential construction and the remainder to non-residential buildings. Interestingly, despite marked differences in the size, economies and degree of industrialization of these countries, and although the total quantity of energy consumed by the New Zealand construction industry is considerably smaller than that of

Table 12.3 Breakdown of annual building industry energy use in the USA and New Zealand for new construction

Building type	Country			
	USA (1967)		NZ (1971–72)	
	(PJ)	(%)	(PJ)	(%)
Residential – new	1 195	32.6	7.8	43.6
Residential – alteration	279	7.6	1.3	7.3
Hotels and motels	74	2.0	0.7	3.9
Hospitals	125	3.4	0.7	3.9
Factories	495	13.5	1.7	9.5
Commercial	586	16.0	2.4	13.4
Schools	467	12.7	1.3	7.3
Others	444	12.1	1.9	10.6
Totals	3 665	100	17.9	100

the USA, the proportions going to the various building types are very similar.

The capital energy required per unit floor area for constructing new buildings is higher in the USA than in New Zealand. For New Zealand residential and commercial construction, the intensities are about 3300 and 4000 MJ/m^2, respectively, compared with figures of 8000 and *c.* 13 000 MJ/m^2 in the USA.

Construction energy requirements account for about 10% of total national energy consumption in both the USA and New Zealand. Energy required for the manufacture of materials and components makes up the largest proportion, amounting to some 70% of the total construction industry energy requirements. Thus the energy embodied in building materials is by far the most important component in the capital energy cost of a building.

Of course, energy use in the construction industry and its impact on the global environment must not be considered in isolation from the other environmental impacts of this sector of the economy. The materials requirements of the construction industry impact on the environment in many different ways during the various stages of the building and related processes.

For example, most of the raw materials needed for building are extracted directly from the earth by surface excavation or underground mining techniques of one kind or another – processes which have considerable and frequently permanent impacts on the environment. Many of the required materials are produced from low grade ores, with the consequences that not only must large quantities of ore be excavated but also large amounts of energy are needed to refine it. The processing of the refined ore into an appropriate building material or component requires further inputs of energy, and the process may itself have undesirable environmental side effects such as the emission of pollutants into the atmosphere or the discharge of toxic wastes into surrounding waterways.

Needless to say, the construction process itself will impact on the building site and on the surrounding environs, as will the completed building, but these are likely to be less influenced by the selection of the building materials than by other considerations such as the use to which the building is put.

Underlying all of this is the environmental impact of the extraction of the fuels needed to produce the energy to drive all of the above processes. The environmental consequences of operations such as coal mining, oil and gas production, and nuclear- and hydro-electricity production do not require further elaboration here.

As will be evident from even this brief summary, the selection of building materials and components which will cause the least environmental impact is a most important task and one which is not at all straightforward.

The energy embodied in some common building materials and their environmental impact potential will now be considered in more detail.

12.4 Building materials – their energy requirements and environmental impact

In this section the processes involved in the production of some common building materials and the factors affecting them will be outlined to give an overview of the range of energy requirements for different varieties of the same material and for the different degrees of refinement. A brief indication of the possible environmental impact of some of these processes will also be given.

The energy requirement, or more precisely the energy coefficient, of a material is the energy required to produce a unit of the material, e.g. the energy per unit mass or per unit volume. It should be noted that the energy coefficient of a material or a process, or even its environmental impact, may not be the same for different countries and different years. This is because of variations in such matters as production sites, raw material quality, climate, processing methods, age and type of plant, efficiency of energy and material use, and basic load requirements. In other words, there is no correct or absolute value for the energy coefficient of a given material and its environmental impact can vary from situation to situation.

There may also be fairly large variations in energy coefficients if the analyses are taken to different levels, i.e. if different system boundaries are drawn. The proportions of coal, oil, gas and electricity making up the total energy requirements for each material will differ between countries and this may further contribute to variations in the energy coefficient. Finally, different conventions and methods of evaluation will contribute to variations in the calculated results. These points should be borne in mind when consulting the work of different authors.

The following building materials and related processes will now be considered in turn:

- cement, concrete and plaster products
- clay products
- glass
- metals
- paints
- plastics
- wood and paper products
- quarrying energy
- transport energy.

12.4.1 Cement, concrete and plaster products

Cement

The manufacture of Portland cement consists of three major stages:

1. raw material processing (quarrying, crushing, drying and grinding limestone and marl);
2. pyroprocessing;
3. finish grinding.

Pyroprocessing takes *c.* 80% of the process energy requirements. Energy coefficients can vary significantly with the manufacturing process, the dry process using 27–40% less energy than the wet. The literature quotes values ranging from 5.7 to 9.4 MJ/kg. Table 12.4 contains 'best estimate' energy coefficients for the USA, the UK and New Zealand.

Concrete

The production of concrete consists of three stages, mixing, forming and curing, the process itself having a low energy requirement by comparison with that needed in the manufacture of the cement and in the quarrying and transport of the aggregates (unless the products are heat cured).

The energy requirement can vary widely depending on the cement : aggregate : sand ratio and the transport requirements. Values range from *c.* 1.6 to 2 MJ/kg in New Zealand, with a stronger concrete usually having a higher energy coefficient.

Table 12.4 Energy coefficients of the main building materials for the USA, the UK and NZ (all coefficients are to IFIAS Level 4 except those indicated with an * which are to Level 2)

Material	Units	Energy coefficients (MJ/unit)		
		USA	UK	NZ
Cement	kg	9.4	5.9	9
Concrete	kg	1	1.7	2
Plaster	kg	5.9	3.2	6.7*
Structural clay	kg	7.2	2	6.9
Glass	kg	8.4	22	31.5
Iron	kg	16	23.8	16*
Steel	kg	35–100	37–47	35–50
Copper	kg	106–66	45	–
Aluminium	kg	211–65	96	130–54*
Lead	kg	26	25	–
Zinc	kg	38	67	–
Paints	l	120	115	–
Plastics – general	kg	–	159	–
Polyethylene	kg	163*	104*	112*
PVC	kg	87*	85*	96*
Polypropylene	kg	157*	171*	–
Polystyrene	kg	140*	96*	–
Finished timber	m^3	3 000	4 480	4 692
Finished timber	kg	4.3	6.4	6.7
Paper	kg	61	42	26*

Plaster

Plaster is made by crushing and dehydrating gypsum and its reported energy coefficient has varied from 3.2 to 7.2 MJ/kg.

One major building product that uses plaster is gypsum board. It is made by mixing plaster with various fillers to produce a paste which is then formed between sheets of paper board and dried. The main energy requirements are for the dehydration of the gypsum and the drying of the boards. The energy coefficient for gypsum board ranges from *c.* 60 to 80 MJ/m^2.

As well as their energy impacts the main environmental degradation associated with this group of materials relates to the disfigurement of the landscape due to the quarrying processes involved, the release of considerable quantities of carbon dioxide into the atmosphere and the production of dust during their manufacture leading to local pollution; in addition, there is the potential for lung and skin problems for those directly involved.

12.4.2 Clay products

The production of structural clay can be divided into:

- quarrying, mining of clay;
- crushing, milling, blending, moistening and de-aerating the clay;
- extruding or forming in a mould;
- drying to remove moisture;
- firing in a kiln, which is the most energy intensive stage.

An average energy coefficient *c.* 7 MJ/kg has been suggested, although values as low as 1.8 (UK) and 3.3 (Australia) have been proposed by some authors.

Apart from the energy requirement the main environmental impact here is associated with the quarrying of raw materials, with its effect on the surrounding landscape, its wildlife and the potential for air pollution.

12.4.3 Glass

The main steps in the manufacturing process are:

- extraction of the raw materials;
- mixing of silica, limestone, sodium carbonate and waste glass;
- melting of raw materials;
- refining the molten glass;
- forming the refined glass;
- finishing.

The melting and refining steps take *c.* 80% of the energy requirement, which can range from *c.* 8 to 30 MJ/kg.

In common with other quarrying processes the extraction of the main raw materials, sand and limestone in this instance, can take a significant toll of the environment.

12.4.4 Metal products

The production of most metals can be considered in four main stages:

- mining and quarrying of the ore, coke, limestone and other raw materials;
- reduction of the ore in a furnace, usually the most energy intensive stage;
- refining and alloying;
- casting, fabricating and machining into the finished products.

The energy requirements are affected to a large extent by the proportion of recycled scrap metal to raw ore, the latter requiring much more energy.

In the UK, for example, for steel with 50% scrap the energy requirement is 47.5 MJ/kg, more than double the requirement of a batch made up entirely of scrap.

Iron and steel are the most extensively used metals in the construction industry. Depending on product and finish, the refining and fabrication stages can be very energy intensive. A semi-finished slab of steel, for example, may have an energy coefficient as low as 28 MJ/kg, whereas a thin sheet of special heat treated alloy steel may have an energy coefficient as high as 210 MJ/kg.

The production of non-ferrous metals, such as lead and zinc, has energy requirements of magnitudes similar to iron and steel; metals such as aluminium and copper tend to be rather higher, as indicated in Table 12.4.

The mining and quarrying of the raw materials needed for the production of ferrous and non-ferrous metals causes large-scale environmental destruction. The primary production processes frequently result in environmental pollution, and, according to Fox and Murrell (1989), some secondary processes are particularly 'eco-sinful' in terms of their toxic by-products. They do go on to say that much can be done to mitigate this type of environmental impact, but make the interesting observation that, 'When these (pollution related) factors are taken into account, any view that metals are environmentally preferable to plastics has to be reconsidered, especially in building applications.'

While some of the raw materials are relatively abundant many are not. Recycling would seem essential for the case of metals, and where it has been tried it has proved to have significant energy advantages.

12.4.5 Paints

Paints have the following constituents: binders (or medium), driers and catalysts (hardeners), pigments, extenders, solvents and thinners and emulsifiers, flattening agents and gelling agents. Depending on the type of paint, and thus the chemical constituents, the energy requirements for manufacture vary considerably, from *c.* 100 to 200 MJ/l.

The likely environmental impact will also vary depending, to some extent, on the solvents and types of pigment used. Solvents constitute a health hazard while the production of metal based pigments gives rise to toxic wastes.

12.4.6 Plastics

The manufacture of plastics can be divided into four major steps:

- acquisition of feedstocks;

- manufacture of monomeric and other inputs to polymerization;
- polymerization;
- product fabrication.

The monomer production and polymerization processes are the most energy and capital intensive steps, amounting to between 80 and 90% of the total process energy requirement for the finished product. The energy requirement also varies widely for the different types of plastics and for different countries.

Table 12.4 gives some indicative values for a range of plastics. It should be noted that most of these are to Level 2 only (see Table 12.2 for explanation of levels) with respect to the production of the 'raw' plastic; though the product fabrication stage is usually much less energy intensive.

The main feedstocks used in the manufacture of plastics are oil, natural gas, coal and salt, their extraction carrying with it a familiar range of environmental hazards. Some of these feedstocks are already being replaced by renewable raw materials and increased effort is being put into the design of systems that will enable the recycling of plastics.

12.4.7 Wood and paper products

The manufacture of wood and paper products starts off with relatively low energy intensity processes – logging, transport and debarking.

In the case of timber production, sawmilling and preservation treatment are also involved – the preservation process being the most energy intensive at *c.* 80% of the total process energy requirement.

For sheet timber products (fibreboards, paperboards, particle boards, veneers, plywood, etc.) pulping, pulp drying and board making are involved, and use over 95% of the energy required. The board making process is the most energy intensive, making up *c.* 50% of the process energy requirement. The energy coefficient of these sheet products can be four or five times that of treated timber.

Following the logging, transport and debarking of the timber, paper production involves pulping, drying and papermaking. The three latter steps consume more than 90% of the process energy requirement. Table 12.4 gives some energy coefficients for finished timber and for paper products.

While, according to Fox and Murrell (1989), wood is 'potentially the ultimate environment-friendly building material', there is still some way to go to achieve that ideal. Many species of timber are approaching extinction, habitats are being destroyed and land surfaces are being eroded. Pearson (1989) recommends the use of sustainable softwoods and some hardwood species from well managed plantations. Concern has also been expressed about the environmental hazards inherent in the chemical treatment of timber and alternative methods have been proposed.

12.4.8 Quarrying energy

In both the USA and New Zealand a large proportion (*c.* 90% by mass and cost) of quarry products go either directly or indirectly to the construction industry. The energy required for extracting natural sand and natural aggregates is *c.* 17 and 19 MJ/tonne, respectively, while that for extracting, crushing and grinding crushed aggregate is *c.* 80–90 MJ/tonne. The deleterious environmental consequences of quarrying have already been alluded to in the above discussion of different materials.

12.4.9 Transport energy

Transport energy requirements often make up a significant proportion of the total in the building industry because of the high weight:value ratio of building materials. These energy requirements average out at *c.* 4.5 MJ/tonne/km for road transport and 0.6 MJ/tonne/km for rail transport in both the USA and New Zealand. In the case of air transport the New Zealand figure is *c.* 50% higher, at 15 MJ/tonne/km, than that of the USA. The environmental impact of the various modes of transport is well described elsewhere and will not be repeated here.

12.4.10 Summary

Baird and Chan (1983) provide a detailed set of tables (35 in total) which list published energy requirement values for all of the main building materials, broken down by analysis method, country, IFIAS level and reference source for the data. Table 12.4 summarizes that data. It lists the energy coefficients of the main building materials for the USA, the UK and New Zealand. Although there are differences in energy coefficients for the three countries which could be as much as 100%, the energy coefficients for the same materials are all of the same order of magnitude.

Difficult as it is to provide reasonably accurate quantification of the energy intensity of these materials, the task pales to insignificance by comparison with what could be involved in attempting to quantify their environmental impact. The Odums (Odum and Odum, 1976) have put forward methods for estimating the energy requirements of processes, human labour and environmental damage, but these have not yet found general application.

Determining the energy coefficients of the building materials is simply the first step. To determine the capital energy requirement of a whole

(a)

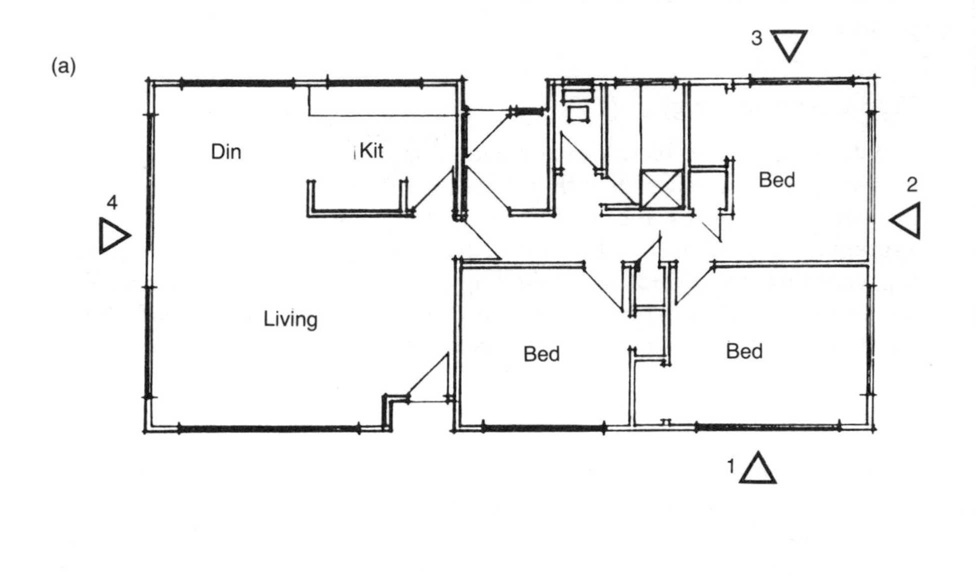

(b)

2420 mm

(c)

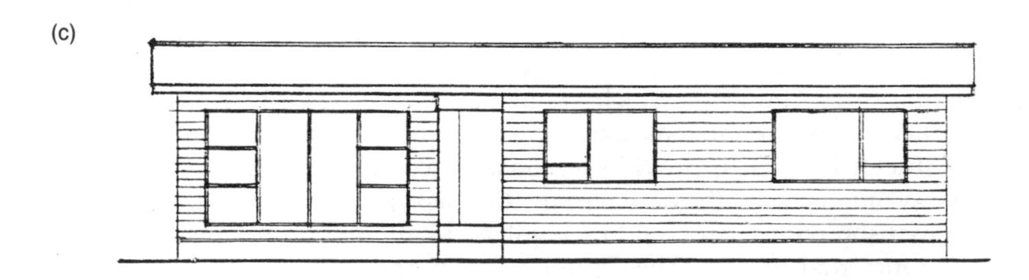

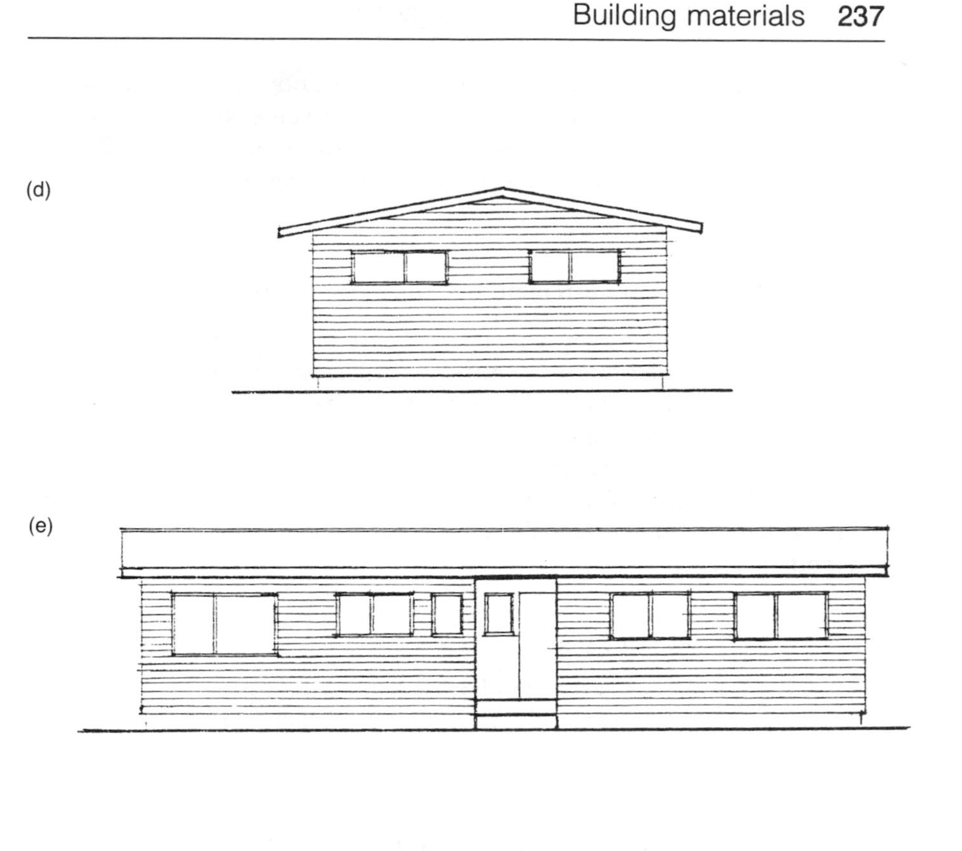

(f)

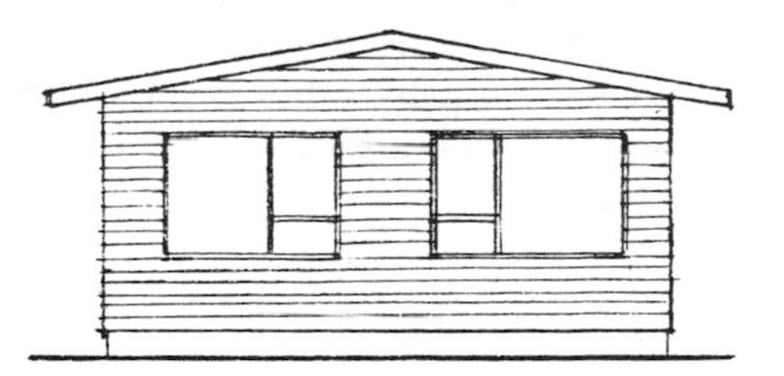

Figure 12.2 New Zealand Building Industry Advisory Council (BIAC) Standard House. (a) Plan view; (b) cross-section; (c) elevation 1; (d) elevation 2; (e) elevation 3; (f) elevation 4. Brief specification of the BIAC Standard House – 94 m^2 floor area; three bedrooms; level site; precast concrete piles; concrete steps; suspended timber floor; weatherboard wall cladding; corrugated iron clad gable roof; timber joinery; particle board floor; gypsum board interior wall lining; sloped ceiling, exposed rafters to dining room and lounge; flat ceiling to other areas; separate shower, bath and laundry; separate WC; 12 electric lights and 16 power points.

building it is obvious that one also needs to know how much of that material is used in its construction and have some knowledge of the construction process and its direct energy requirements. A method of deriving the energy requirements of whole buildings will now be demonstrated.

12.5 Capital energy requirement of a house – a case study

In this section a sample capital energy requirement estimate for a single-storey timber-framed house is introduced. Using this as a reference the effects of variation in the building materials used on the capital energy requirement are estimated.

It should be noted that the standard house introduced here, although typical of houses in New Zealand, may not be typical of those in the USA or the UK. It serves only as an illustration of the principles and methods involved.

12.5.1 The BIAC Standard House

The building selected is the standard house of the former New Zealand Building Industry Advisory Council (BIAC), which is fairly typical of a large number of houses in New Zealand. The brief specification for the 94 m^2 house, together with its plan, cross section and elevations, is given in Figure 12.2.

The procedure for estimating the capital energy requirement outlined below has three main steps. First, the determination of the energy coefficients for the relevant building materials; then the estimation of the gross energy requirement (GER) of the various components; and finally, the summing of the component GERs to give the total capital energy requirement for the house.

12.5.2 Energy coefficients of building materials

While the values presented in Table 12.4 give an indication of the GERs for the main building materials, it is by no means an exhaustive list – its purpose was to give a broad summary of the figures one might expect in practice and to enable international comparisons to be made.

Table 12.5 is more indicative of the kind of information necessary to enable capital energy requirements to be estimated with known accuracy. The table lists the New Zealand energy coefficients for a wide range of building materials and activities associated with the building process. They are all given in MJ per unit of the corresponding material and their level is specified (most are to IFIAS Level 4 in this instance; Baird and Chan, 1983).

Table 12.5 New Zealand energy coefficients for a range of building materials and related activities (Source: Baird and Chan, 1983; Items 7a and 7b from Honey and Buchanan, 1992)

No.	Material/work	Unit	MJ/unit	IFIAS level
1	Profits	No	0	0
2	Preliminaries	$	39.5	4
3	Administration	$	22.6	4
4	Earthwork	m^3	100	1
5	Labour	No	0	0
6	Sitework (oil)	MJ	1	1
7	Timber: rough	m^3	848	4
7a	Timber: air-dry, treated	m^3	1 200	4
7b	Timber: glulam	m^3	4 500	4
8	Timber: milled	m^3	4 692	4
9	Timber: formwork	m^3	283	4
10	Hardboard	m^3	20 626	4
11	Softwood	m^3	15 469	4
12	Particleboard	m^3	12 892	4
13	Plywood	m^3	9 439	4
14	Veneer	m^3	11 208	4
15	Wallpaper	m^2	14.92	4
16	Building paper	m^2	7.46	4
17	Cement	kg	8.98	4
18	Concrete: precast	m^3	4 780	4
19	Concrete: in situ	m^3	3 840	4
20	Lime mortar 1:2	m^3	2 500	4
21	Cement mortar 1:2	m^3	5 980	4
22	Structural clay	kg	6.9	4
23	Other clay	kg	200	2
24	Plaster: solid	kg	6.7	2
25	Plaster: fibrous	kg	6.7	2
26	Gib board	m^3	5 000	2
27	Asbestos cement	kg	8.2	4
28	Asbestos: others	kg	8.2	4
29	Asphalt felts	kg	31	4
30	Bitumen felt	kg	38	4
31	Glass	kg	31.5	4
32	Steel: general	kg	35	4
33	Steel: rods	kg	35	4
34	Steel: sections	kg	59	4
35	Galvanized iron	kg	37	4
36	Steel pipes	kg	57	4
37	Non-ferrous metals	kg	0	0
38	Aluminium: general	kg	129.5	2
39	Aluminium sheets	kg	145	2
40	Aluminium extrusion	kg	145	2
41	Aluminium foil	kg	154	2
42	Copper	kg	45.9	4
43	Zinc	kg	68.4	4
44	Lead	kg	25.2	4

Table 12.5 *Continued*

No.	Material/work	Unit	MJ/unit	IFIAS level
45	Plastics: general	kg	160	4
46	Polyethylene	kg	112	2
47	Polystyrene	kg	100	2
48	PVC	kg	96	2
49	Polypropylene	kg	175	2
50	Paints: general	m^2	15	4
51	Paints: water-soluble	kg	7.5	4
52	Paints: emulsion	m^2	10	4
53	Paints: oil-based	m^2	12	4
54	Electrical work	$	39.5	4
55	Wiring	m	4.38	4
56	Electric equipment	$	46.7	4
57	Electric range	No	6 456	4
58	Aggregate	kg	0.3	4
59	Masonry, stone	kg	0.3	4
60	Sand	kg	0.04	4
61	Rubber: synthetic	kg	148	4
62	Insulation: fibre	kg	23	4
63	Fibreglass batts	kg	150	4
64	Brass	kg	49.3	4
65	Asphalt: strip, shingle	m^2	280	4
66	Asphalt: surface rolled	m^2	85	4
67	Chip-seal, pavement	m^2	8.41	1
68	Lime: hydrated	kg	10.4	4
69	Quicklime	kg	7.4	4
70	Sitepower	MJ	1	1
71	Sitepower	$	300	1
72	Transport: road–30 km	Tonne	114	1
73	Transport: road–50 km	Tonne	190	1
74	Transport: road–100 km	Tonne	230	1
75	Transport: rail–200 km	Tonne	146	1
76	Transport: rail–500 km	Tonne	365	1
77	Transport: general	$	35	1
78	Nil	No	0	0
79	Nil	No	0	0
80	Nil	No	0	0

12.5.3 Gross energy requirements

Having specified the energy coefficients the next step is to apply these to the quantities of building materials used in the construction of the house. This can be done in several ways, but will usually involve the subdivision of the building into its various components. Here components such as floor, wall, roof, etc., have been used so that comparisons between different materials and methods of construction will be straightforward.

As an example of the computation involved for, say, the floor construc-

Table 12.6 Computation of the GER for the floor construction of the BIAC Standard House

Floor construction – list of materials	Unit	Net amount	Coefft. (MJ/unit)	Waste (%)	Energy (MJ)
Excavation and ground work	m³	1	100	0	100
Precast concrete piles	No	36			
and footings (600 × 200 × 200 on 300 × 300 × 100)	m³	1.18	4 780	10	6 200
Bearers (100 × 100)	m	56			
	m³	0.56	4 690	10	2 890
Floor joists (150 × 50 at	m	245			
450 crs)	m³	1.84	4 690	10	9 500
Particle board flooring,	m²	95			
high density 20 mm	m³	1.9	12 890	20	29 390
Nails and fasteners	kg	12	35	5	440
Concrete steps	m³	3	3 840	10	12 670
Reinforced steel	kg	20	35	10	770
Total gross energy requirement (MJ)					61 960

Table 12.7 Summary of the component GERs of the BIAC Standard House and its total GER

Components	Gross energy requirement (GJ)	(%)
Floor	62	17.4
Walls	50	14.0
Roof	66	18.5
Joinery	62	17.4
Plumbing	28	7.8
Electrical	5	1.4
Finishes	57	16.0
Insulation	16	4.5
Preliminaries	11	3.0
Total GER	357	100.0
GER/m² floor area	3.8	

tion of the standard house a detailed list of materials with their respective energy coefficients, net quantities and estimated waste, together with the corresponding gross energy requirement, is given in Table 12.6.

Table 12.7 summarizes the data for the major components of the house as a whole. The GER values, which are to IFIAS Level 4, include material transport to the site, on-site energy requirements and material wastes.

As can be seen from Table 12.7, the total GER or capital energy requirement of the BIAC House amounts to 357 GJ, or 3.8 GJ/m² of floor area,

Table 12.8 Gross energy requirements of some common New Zealand domestic-scale building assemblies (Baird and Chan, 1983)

Type of assembly	GER per unit area (MJ/m^2)
Walls	
Timber frame with weatherboard cladding	198
Timber frame with brick veneer cladding	1 284
Concrete block wall	755
Roofs	
Galvanized iron roof	508
Concrete tile roof	176
Floors	
Suspended timber floor	733
Concrete slab on ground	1 014

with the floor, walls, roof, joinery and finishes making up nearly 85% of the total.

12.6 The effect of using different building materials and components on capital energy cost

Having estimated the capital energy cost of the house described in the previous section it is now of interest to gauge the effect of the use of alternative building materials and components. The BIAC Standard House had a suspended timber floor, weatherboard walls and a galvanized iron roof. The most popular alternatives to these in New Zealand are the concrete slab floor, walls of concrete block or of timber framing with a brick veneer, and concrete tiled roofs.

In this section the energy requirements of these major components are compared and the implications of their use in different combinations, for the capital energy cost of the whole house, are explored.

12.6.1 Comparison of building component GERs

Baird and Chan (1983) estimated the gross energy requirement per unit area of each of the building components noted above, namely, three wall types, two roof assemblies and two types of floor construction. Their results are summarized in Table 12.8.

As can be seen from Table 12.8, the greatest variation occurred for the three wall types, with the GER of the brick veneer construction over six times that of the weatherboard, due to the high energy requirement of brick. GER values of this order of magnitude were also found by Stein *et al.* (1976) in the USA. In terms of whole house energy capital costs the use of brick veneer would result in a 30% increase. The GER of the concrete block wall is almost exactly midway between that of weatherboard and brick veneer.

The difference between the two roof assemblies is less dramatic than for the walls, but is still relatively significant with galvanized iron nearly three times the GER of concrete tile at the scale of domestic construction. The magnitude of this difference is likely to be less in the case of the roofs of larger buildings due to differences in the supporting structure.

The two floor constructions exhibited the least difference, and even this is likely to reduce as cement manufacturers convert to the dry process (the figures quoted being based on the more energy intensive wet process of cement production).

From this brief review, it can be seen that there is a considerable range in the energy requirement per unit component area, even for some fairly basic and relatively similar constructions. It is clear that the choice of building material will have a significant effect on the total capital energy requirement of a building. The extent of that effect will now be explored and quantified.

12.6.2 Effect of choice of building materials on the capital energy cost of the BIAC Standard House

To examine the effect of differences in materials on the energy required for construction, the capital energy cost of the New Zealand BIAC Standard House was estimated for the range of wall, roof and floor assemblies described in the previous section. The results are presented in Table 12.9 in terms of capital energy cost per unit floor area.

Table 12.9 Variation in capital energy cost (GER) of the Standard House with different combinations of building assemblies

Building assembly combinations			House GER
Floor	Wall	Roof	(GJ/m² of floor area)
Timber	Weatherboard	Concrete tile	3.4
Timber	Concrete block	Concrete tile	3.7
Timber	Weatherboard	Galvanized iron	3.8
Concrete	Weatherboard	Concrete tile	3.9
Timber	Concrete block	Galvanized iron	4.2
Concrete	Concrete block	Concrete tile	4.3
Concrete	Weatherboard	Galvanized iron	4.4
Timber	Brick veneer	Concrete tile	4.5
Concrete	Concrete block	Galvanized iron	4.7
Timber	Brick veneer	Galvanized iron	4.9
Concrete	Brick veneer	Concrete tile	5.1
Concrete	Brick veneer	Galvanized iron	5.5

As would be expected from the previous discussion, the lowest energy option is the house with weatherboard wall cladding, suspended timber floor and tile roofing, while the highest energy option has brick veneer walls with a concrete slab on ground floor and galvanized iron roofing. The difference is about 2100 MJ.

Thus, the capital energy cost of a house can be *c*. 60% more simply due to differences in commonly available construction materials. This is equivalent to several years' annual operating energy consumption. The selection of basic building materials for housing can therefore have significant energy implications.

12.7 International comparison of capital energy requirements for buildings

Just as variations in materials and construction methods may be expected to produce different GER values within a particular country these, and other, factors will result in differences between countries and between building types. The energy requirement shows a considerable range, even for the same building type. This is illustrated in Table 12.10 where it will be seen that domestic buildings in the UK and the USA appear to be more energy intensive than those in Australia and New Zealand. These variations are most likely due to the different building practices and material energy coefficients in these four countries.

Table 12.10 also notes the ratio of the capital to the annual operating energy requirements. This ratio provides an index by which different building designs may be judged and alternative investments in energy

Table 12.10 Comparison of GERs and ratio of capital to operating energy requirements for a range of domestic building types

House type	Floor area (m^2)	GER (GJ/m^2)	Capital/operating energy (years)	Reference
UK three bedroom semi-detached	100	4.0	4.8	Brown and Stellon, 1974
UK standard mid-terrace house	93	7.0	3.1	Markus and Slessor, 1976
USA single-family dwelling	–	8.0	6	Stein *et al.*, 1976
Australian single-storey	144	3.6	6	Hill, 1978
NZ timber frame house (BIAC)	94	3.8	9.8	Baird and Chan, 1983
NZ timber frame house (NZIV)	93	3.6	9	Baird and Chan, 1983

Table 12.11 Capital energy requirements for a range of building types (Source: Stein *et al.*, 1976)

Building type	Capital energy requirement (GJ/m^2 floor area)
Warehouses	6.3
Residential family	8.0
Shops and restaurants	10.7
Industrial buildings	11.0
Hotels and motels	12.8
Religious buildings	14.3
Educational buildings	15.7
Office buildings	18.6
Hospitals	19.6
Libraries and museums	19.9
Laboratories	23.6

efficiency compared. For non-domestic buildings the ratio will be important for examining the energy cost effectiveness of heating, ventilating and air-conditioning systems. The ratio can also be used to gauge the effectiveness of thermal insulating systems.

The data presented in Table 12.10 is consistent with the assumption that the value of the ratio will be lower in a location with a more severe climate, if other factors are equal, because of the higher operating energy requirement. Possibly most notable of all is the relatively short time (between three and 10 years) it takes for an amount of energy equivalent to the initial capital energy requirement to be consumed in houses designed and operated to the norms of the 1970s.

Our knowledge of the capital energy requirements of building types other than domestic dwellings is almost entirely due to the work of Stein *et al.* (1976). Table 12.11 is based on their work and gives the average capital energy requirement of 11 different building categories, ranked in order of intensity per square metre of floor area.

Apart from warehouses, it can be seen that every other category of building has a significantly higher energy capital cost than housing. Not surprisingly, laboratories were found to have a relatively high energy requirement per unit floor area, nearly three times that of houses, while schools, offices and hospital buildings were around double. This serves as a reminder, given that most researchers have concentrated on housing, that the potential for energy capital cost savings is likely to be even greater in commercial, institutional and industrial buildings.

In summary, the capital energy requirements of houses vary from *c.* 3 to 8 MJ/m^2 of floor area, depending on the type and materials of construction. A house with basically the same structure, for example, can require up to 60% more capital energy as a result of the designer's choice of building materials. The capital energy requirement for housing in the

USA and the UK appears to be higher than in Australia and New Zealand due to differences in building practices and materials. In the case of current housing design the ratio of capital to operating energy requirement can range from about three to 10 years. This ratio is important for gauging the energy cost effectiveness of 'energy saving' construction and operating systems. The situation for non-domestic buildings is less well defined as most research has concentrated on the relatively less complex house types, though Szokolay (1980) suggested a rule of thumb that gave office buildings five times and industrial buildings 10 times the capital energy cost of domestic dwellings.

12.8 Conclusions

It will have become increasingly clear from the information given in this chapter that the selection of building materials from the point of view of energy efficiency and the reduction of global environmental impact is not a trivial task. Not only are the methods which would enable the conscientious building provider to take account of these matters in their infancy, but the data to enable them to be applied are not yet readily and generally available. Only a relatively small amount of research has been carried out in this field, much of it a decade or more ago, and it is timely to refocus on the issues involved.

It is with some disquiet that one realizes that much of the current building stock and the infrastructure to support it were laid down during the 1950s and 60s, the era of so called 'cheap' energy, and, given their normal expected lifetime, they will be with us for some considerable time yet. In addition, one is left with the impression that most building providers over the last 1000 years or so have treated the global environment as an infinite resource. Indeed, the Vales (Vale and Vale, 1991) go so far as to suggest that the city represents the most extreme example of not working with nature that people have yet produced on the planet.

Given their inevitable focus on the activities of the building site itself it may be understandable that the building providers were lulled into believing that what they were doing was improving the environment. In practice, regrettably, the vast majority of the deleterious aspects of their activities remained well hidden from general view in mines, quarries, processing plants and factories remote from the building site, or else manifested themselves later in the energy and maintenance requirements of operating the buildings.

The growing realization of the real impact of buildings on energy and environmental resources, brought about by investigation of their role in relation to the greenhouse effect, must give fresh impetus for a concerted effort to further develop the methods to enable building providers to take these matters into account.

One aim of this chapter was to summarize the present status of work in this field. As will be evident, most of the effort has gone into methodologies related to the assessment of energy capital costs, very little into the quantification of environmental costs and a minimal amount into their overall integration. These matters, together with the need for a change in attitude on the part of building providers and for further research in this area, will now be summarized briefly.

Of course, current economic philosophies would imply that competition and the price mechanism should give sufficient information to enable rational selection of building materials on the basis of cost. Unfortunately, this is not the case. The full environmental costs of the production of energy and materials do not yet appear in the accountant's balance sheet.

The best advice on offer may well be to ensure that one uses the least amount of building materials, consistent with the minimization of life cycle energy and resource costs; yet another interpretation of 'less is more'. Good as that may be it is nowhere near satisfactory in the light of the potential disaster faced by the global environment. Hence the identification of materials with a low environmental impact and a low use of energy is of paramount importance to us all.

12.8.1 Building materials and their environmental impact

None of the conventional raw materials used in the construction of buildings is without some impact on the environment. That impact can be either direct – on the local environment in the vicinity of the mine, quarry or plantation where the raw material is extracted and at the production facility – or indirect in terms of the energy needed to process, transport and refine it.

The amounts of materials and how they are used is also of importance in determining the overall impact on the global environment. A high energy insulating material, for example, may enable even greater energy savings in the operation of a building. 'Environmentally difficult' plastics with the ability to be recycled at the end of each use may be preferable in the long run to the use of relatively lower impact materials which can only be discarded.

Another of the key issues in this respect is whether the material is capable of being produced sustainably. This is clearly not the case for metals and several of the feedstocks for the plastics and paint industries. Meadows *et al.* (1972) predicted that known reserves of metals such as aluminium, copper, lead and zinc would be exhausted within a few decades at present rates of use (iron was predicted to last around a couple of centuries). Clearly, the rate of use far exceeds their rate of production in the earth's geological cycle. A similar picture emerges for oil, gas and coal. Even the raw materials needed for the manufacture of concrete, cement, plaster and glass, namely, limestone, sand, gypsum, salt, aggregate and so on, while relatively plentiful, are not infinite. Their extraction can impact badly on the surrounding area and the associated transport requirements add their own particular impacts.

While the mineral accretion and solar-generated construction materials developments described by Hilbertz (1991) hold out some hopes for the future, currently only wood appears to have the potential to be produced in line with present day concepts of sustainability. This involves the management of forests as a sustainable resource, matched to the needs of the environment as well as those of the timber processing industries.

As will be evident, there is no easy gauge of environmental impact that can be applied to the variety of materials used in buildings and it seems that most of them are far from being environmentally benign. Despite this, some commentators offer advice in terms of the desirability of, for example, using local materials to reduce transport requirements, using timber products (but not tropical hardwoods) and so on. Pearson (1989) is one of the more explicit, recommending that materials meet as many as possible of the following criteria if they are to be considered environmentally sound: renewable and abundant; non-polluting; energy efficient; durable; equitable; recyclable.

12.8.2 Building materials and their energy requirements

As indicated in earlier sections of this chapter, it is feasible to estimate the energy embodied in a large range of building materials and use this to make decisions about their selection. If only because of the limited reserves of conventional fuels and the environmental and other impacts associated with their extraction and use, not the least of which is the greenhouse effect, it is important that the effort be made to reduce this form of energy use.

Apart from using low energy materials, designing adaptable buildings to last for hundreds rather than scores of years has the potential to save vast amounts of capital energy (Storey and Baird, 1993). Needless to say, the choice of materials and their detailing will need to be compatible with this concept to avoid the energy cost penalties of frequent replacement of components.

It should also be noted that despite the nomenclature, the energy capital cost of a building is unlikely to be able to be cashed in at the end of its useful life, unless it has been designed such that its components can be readily dismantled and its materials recycled – properties which are not a characteristic of the current building stock. Clearly, if this capital energy is to be recoverable then building components must be designed for ease of disassembly.

Capital energy costs must always be considered in the context of the total energy requirement over the lifetime of the building. In other words, the operating energy requirement must also be estimated and the influence of alternative materials selections compared. Energy 'costing' does not need the sophistication of discounted cash flow and related methods. A simple payback calculation will normally be sufficient to allow assessment of alternatives.

In the case of house thermal insulation, for example, Hill (1978) has shown that the payback times, in terms of savings in heating energy consumption, can be very short indeed. Even for the relatively mild climate of Melbourne, Hill's analysis of six different insulating materials indicated that their energy capital cost would be recouped within a single heating season.

The examples of section 12.6 give ample demonstration of the effect of the use of different materials on the energy requirement of a range of building components and the influence this can have on the total capital energy needed to build a simple house. Baird and Chan (1983) provided a further example of the type of comparison that is feasible. In what they describe as a tentative analysis, the energy requirements of timber window frames were about one-tenth the figure for comparable aluminium frames.

Finally, in relation to building materials, it is worth reiterating that the energy requirement will often be a good indicator of the environmental impact of a particular material (see, for example, Howard, 1991; Peet, 1991). While it does not tell the whole story, the fact that so many of the indirect processes – in particular the extraction, transportation and refining of the raw materials – involved in producing a building are both energy and resource intensive carries broader implications for the environment than just the consumption of energy.

12.8.3 The role of the building provider

It will have become evident in the course of this chapter that the responsibilities of those involved in the provision of buildings must extend well beyond the boundaries of the immediate building site. Not only must the needs of those who use the building be satisfied, a difficult enough task at the best of times, but the need for sustainability of the global environment must also be considered.

To some extent, those who provide buildings are already accustomed to taking into account the likely effect of their products on the wider public, but usually only in terms of their visual impact. Any further consideration of environmental impact would normally be confined to local effects such as changes to pedestrian level wind characteristics, emissions from boiler flues and cooling towers, noise from ventilation fans and increased traffic levels in the vicinity of the building.

However, a fairly major change in attitude will be needed before an awareness of the broader implications of building becomes as commonplace as the above considerations. There are already signs that such a change is taking place, at least in relation to specific issues such as CFCs. Some groups of architects are in the process of drafting environmental policies to guide their members in this process (NZIA, 1992), and the Building Research Establishment has developed an environmental assessment scheme for various building types (see, for example, Baldwin *et al.*, 1990). As architects have probably always considered themselves as improvers of the environment this transformation will be difficult to achieve. The fact that the production of most building materials has a measurable energy cost and a recognizable environmental impact will come as an uncomfortable shock to most.

Nevertheless, an increasingly knowledgeable clientele may be expected to accelerate this change by requiring new buildings to be designed with some cognisance of the broader environment, and increasingly sophisticated tenants are likely to shun buildings which are clearly environmentally irresponsible.

It will also have been evident from the material presented that the current level of information and guidance on this aspect of building design is severely limited. An attempt has been made to summarize the data and procedures of energy analysis relevant to buildings and to outline the processes and likely environmental impacts involved in producing a range of building materials. It is hoped that these will provide at least interim guidance for the environmentally aware designer.

12.8.4 Priorities for future work in this field

Before one can routinely evaluate the capital energy requirements and environmental impacts of different buildings and building components the information available to the designer requires substantial upgrading. While there is some energy data available for domestic construction in several countries, even that has resulted mainly from one-off research exercises which have not been updated to take account of changing processes. The energy data for commercial buildings is even more sparse, as is quantitative information on the environmental impact of building materials, making the task particularly daunting for the environmentally conscientious designer.

Even those pioneering practitioners of green architecture, the Vales (Vale and Vale, 1992), in their recent design of some cottages in North Sheffield, had as their main goal the reduction of operating energy requirements, this being seen, quite correctly, as a key issue in the reduction of both fossil fuel use and the related environmental impact over the lifetime of the building. Having resolved that issue within the constraints set, environmental concerns related to the selection of the building materials were then given careful consideration. For example, their solution involved the use of timbers from sustainable sources, non-CFC insulating materials, low-glue building boards, jute-based linoleum and water-based emulsion paints.

Energy capital costs were also given careful consideration. According to the designers, 'A serious attempt was made to use materials that did not require large amounts of manufactured energy, producing a "wholemeal" architecture of largely unrefined "low energy" materials such as timber for floors and roof, and concrete for ground floor, roof tiles, internal partitions and facing bricks' (concrete facing bricks using less energy to manufacture than fired clay bricks). For the immediate future, this kind of approach, based on a well-informed feel for the energy content and environmental impact of building materials, with a dash of common sense and intuition, is probably all that is available for the building provider who takes these matters seriously. Some more detailed information on the energy content of materials is also available and this may be used to obtain broad estimates of energy capital costs for purposes of comparison of alternative designs and with energy operating costs. Some Australian and New Zealand based authorities have given consideration to the incorporation of capital energy costs in their latest building standards but have not yet done so.

In the case of environmental impact, energy content gives a first approximation, but must be moderated in the light of the direct impact of the extraction, manufacturing and other processes involved. Pearson (1989), in the context of the personal health of the building user and the environmental health of the planet, calls for a reclassification of building materials taking these matters into account.

Ideally, what is needed is a database containing estimates of the range and distribution of Level 4 GERs of building materials and components, manufactured and assembled in different ways, together with a comparative assessment of their overall environmental impact, plus consistent methodologies for the use of such a database to enable realistic comparisons to be made of the effects of using different materials.

In the past such a concept would have been unrealistic, but the development of input–output analysis techniques, the availability of significant computer power, plus the traditional methods of the quantity surveyor, not to mention the motivation provided by the new wave of environmental consciousness, has made this an attainable goal. However, without the database described above, evaluation of the energy and environmental cost-benefit of alternative design options will not be possible. The time is long overdue for a more sustained research effort in this direction. Perhaps this is another worthwhile challenge for the member countries of the International Energy Agency.

Acknowledgements

It is a great pleasure to acknowledge the assistance of colleagues at the Victoria University School of Architecture and the Energy Research Group, in particular Harry Bruhns, Michael Donn and John Storey, for their constructive comments on the early drafts. I should also like to acknowledge the painstaking work of former Research Fellow Chan Seong Aun and the funding of the former New Zealand Energy Research and Development Committee which enabled the creation of the energy coefficients used in this chapter.

References

Baird, G. and Chan, S. A. (1983) Energy Cost of Houses and Light Construction Buildings and Remodelling of Existing Houses. Report No. 76. New Zealand Energy Research and Development Committee, University of Auckland, Auckland.

Baldwin, R. *et al.* (1990) BREEAM 1/90 – an environmental assessment for new office designs. Building Research Establishment, Watford, UK.

Ballantyne, E. R. (1976) *Energy Cost of Dwellings*. CSIRO Division of Building Research, Reprint 704, Melbourne.

Brown, G. and Stellan, P. (1974) The material account. *Built Environment*, **3** (8), 415–17.

Cole, R. J. and Rousseau, D. (1992) Environmental auditing for building construction. *Building and Environment*, **27** (1), 23–30.

Connaughton, J. N. (1992) Real low-energy buildings: the energy costs of materials, in *Energy Efficient Building – A Design Guide* (eds S. Roaf and M. Hancock), Blackwell Scientific Publications, Oxford.

Chapman, P. F. (1975) The energy cost of materials. *Energy Policy*, **47**, 47–57.

Fox, A. and Murrell, R. (1989) *Green Design*. Architecture Design and Technology Press, London.

Haseltine, B. A. (1975) Comparison of energy requirements for building materials and structures. *The Structural Engineer*, **53** (9), 357.

Hilbertz, W. H. (1991) Solar-generated construction material from sea water to mitigate global warming. *Building Research and Information*, **19** (4), 242–55.

Hill, R. K. (1978) *Gross Energy Requirements of Building Materials*. Proceedings of the conference on 'Energy Conservation in the Built Environment', March 1978. Department of the Environment, Housing and Community Development, Sydney.

Honey, B. G. and Buchanan, A. H. (1992) Environmental Impacts of the New Zealand Building Industry. Research Report 92–2. Department of Civil Engineering, University of Canterbury, Christchurch.

Howard, N. (1991) Energy in balance. *Building Services*, **13** (5), 36–8.

IFIAS (1974) Energy Analysis Workshop on Methodology and Convention. International Federation of Institutes for Advanced Study, Workshop Report No. 6, Stockholm.

IFIAS (1975) Workshop on Energy Analysis and Economics. International Federation of Institutes for Advanced Study, Workshop Report No. 9, Stockholm.

Kegel, R. A. (1975) The energy intensity of building materials. *Heating, Piping, Air Conditioning*, **47** (6), 37–41.

Kohler, N. (1987) *Energy Consumption and Pollution of Building Construction*. Proceedings of the 3rd International Conference on Building Energy Management, Lausanne. Part II (ed. A. Faist, E. Fernandes and R. Sagelsdorff), pp. 233–40. Ecole Polytechnique Fédéral de Lausanne.

Mackillop, A. (1972) Low energy housing. *The Ecologist*, November, 4–10.

Markus, T. and Slessor, M. (1976) Housing Energy Economics Pilot Study, OP 76111. Scottish Development Department Research Project, University of Strathclyde, Glasgow.

Meadows, D. H. *et al.* (1972) *The Limits to Growth*. Universe Books, New York.

Ministry for the Environment (1990) *Responding to Climatic Change*. Wellington, New Zealand.

Ministry of Commerce (1990) *An Economic Analysis of the Issues and Options for Reducing Greenhouse Gas Emissions*. Wellington, New Zealand.

NZIA (1992) *Environmental Policy Position Papers*. Institute of Architects, Wellington, New Zealand.

Odum, H. T. and Odum, E. C. (1976) *Energy Basis for Man and Nature*. McGraw-Hill, New York.

Pearson, D. (1989) *The Natural House Book*. William Collins, Sydney.

Peet, J. (1991) Ecological economics, energy and sustainable development. *Perspectives in Energy*, **1**, 219–34.

Pick, H. J. and Becker, P. E. (1975) *Direct and Indirect Uses of Energy and Materials in Engineering and Construction. Applied Energy 1*. Applied Science Publishers, London.

Stein, R. G. *et al.* (1976) *Energy Use for Building Construction*. CAC, University of Illinois and R. G. Stein and Associates, New York, EDRA Contract EY076-S-02-2791.

Stein, R. G. *et al.* (1977) *Energy Use for Building Construction – Supplement*, CAC, University of Illinois and R. G. Stein and Associates, New York.

Stein, R. G. (1977a) Energy cost of building construction. *Energy and Buildings*, **1**, 27.

Stein, R. G. (1977b) *Architecture and Energy*. Anchor Press/Doubleday, New York.

Stein, R. G. *et al.* (1980) *Handbook of Energy Use for Building Construction*. US Department of Energy, DOE/Cs/20220-1.

Szokolay, S. V. (1980) *Environmental Science Handbook*. The Construction Press, Lancaster, UK.

Storey, J. B. and Baird, G. (1993) *Sustainability in the City*. Proceedings of the Annual Conference of the Institution of Professional Engineers, IPENZ, Wellington.

Vale, B. and Vale, R. (1991) *Green Architecture*. Thames and Hudson, London.

Vale, R. and Vale, B. (1992) Low energy cottages. *Building Services*, **14** (2), 19–22.

Wright, C. and Gardiner, P. (1979) Energy and housing. *Built Environment*, **5** (4), 287–97.

Index